The Animal Activist's Handbook

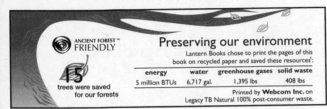

The
Animal Activist's
Handbook

Maximizing Our Positive Impact in Today's World

Matt Ball and Bruce Friedrich

With a Foreword by Ingrid Newkirk

Lantern Books • New York

A Division of Booklight Inc.

2009
Lantern Books
128 Second Place, Garden Suite
Brooklyn, NY 11231

Printed in Canada

Library of Congress Cataloging-in-Publication Data

Ball, Matt.
The animal activist's handbook : maximizing our positive impact in today's world / Matt Ball and Bruce Friedrich.
    p. cm.
 ISBN-13: 978-1-59056-120-1 (alk. paper)
 ISBN-10: 1-59056-120-1 (alk. paper)
  1. Animal welfare–United States–Handbooks, manuals, etc. I. Friedrich, Bruce. II. Title.
 HV4764.B35 2009
 179'.3–dc22
                                    2008032771

ACKNOWLEDGMENTS

Many people gave us feedback on this project. We'd like to offer special thanks to Eileen Botti, Jon Camp, Steve Kaufman, Gaverick Matheny, Ingrid Newkirk, Lauren Panos, and Debra Probert for their edits and comments on this manuscript; Peter Singer, whose writings have been a continuing source of inspiration and insight; and Martin Rowe for seeing the potential for this book, waiting over a year for the first manuscript, and offering extensive and useful suggestions for improving the original manuscript—if he spent this much time on all of the books he published, he'd need nine lives. Bruce would also like to thank his wife, Alka Chandna, who never tires of insightfully and incisively discussing the topics in this book, and Gracie, the world's most perfect cat—and an excellent reminder that other animals are interesting individuals in their own right. He'd also like to thank everyone at PETA (the best place in the world to "work"), whose competence and dedication keep him sane and inspired. Matt would like to thank Jack Norris, without whom Matt would never have become the activist and person he is today; and, as always, deepest gratitude and love to Anne and Ellen Green.

This book actually started as a collection of writings by Bruce and Matt. The significant efforts of Anne Green and Martin Ball edited these pieces together into what we hope is a coherent whole. For references, links, and more information, please visit animaladvocacybook.com.

CONTENTS

Ingrid Newkirk

If you have picked this book up, good for you. It can only mean that you care about living a meaningful life, and what can beat that? Nothing!

The odd thing is how so many people are totally oblivious to the possibilities that exist for them, the "go for its" that dangle right there at their fingertips. They never realize how easy it is to have great experiences that you can look back upon and feel good about. Instead, their lives go dashing by, filled with pointless activity and time wasted on purposeless rubbish. These people may only wake up to the fact that they have squandered their years when there are no years left! That's when the adage "better late than never" no longer applies. They are out of here, opportunities lost. No second chance.

The Chinese philosopher Lin Yutan described wisdom as "reality plus idealism plus humor," and that, in turn, describes the authors of this book, Bruce Friedrich and Matt Ball. Their idea of total hell would be to lead that sort of *I, Robot* or *Life of Brian* existence, blindly buying whatever advertisers suggest you buy, and doing, thinking, and saying what your parents or friends do and think and say. Matt and Bruce have carefully carved out their lives for themselves.

The authors are real movers and shakers and doers (or, to paraphrase a cartoon I saw recently, they would never let some little thing like not being able to find their pants stand in the way of their dreams). They seize every waking moment and claim it as their own. If someone yelled, "Get a life!" to these guys, each of them could quite honestly reply, without any hesitation or fibbing, "Got one!"

Bruce and Matt made their share of mistakes along the way, but, by spelling them out, they spare you the trouble of making them yourself! Not that there's much harm in making your own, for, as George Bernard Shaw once said, "A life spent making mistakes is not only more honorable but more useful than a life spent doing nothing."

One of their many useful suggestions is learning how to connect very easily with countless strangers. Bruce and Matt pretty much started a twenty-first-century revolution by reviving the art of pamphleteering, something that was used to great advantage by the Suffragettes to secure the vote, by Abolitionists to overthrow human slavery, and by every social cause that has existed since *homo sapiens* learned to print. Pamphleteering deserves some of the credit for the positive changes in the world regarding civil rights, women's rights, disability rights, gay rights, and animal rights. In this book, the authors nail exactly how to do it and how much impact and enjoyment you can have in the process.

When I wrote this foreword, I had just finished putting together *One Can Make a Difference*, a book of essays about people who have done exciting and positive things with their lives. Whether they are hugely famous—like the Dalai Lama, Paul McCartney, and Russell Simmons—or complete unknowns, every person I chose made the decision to put their voice, their talent, their values, their time, their freedom, their interests, and their heart to good use. The book you are holding in your hand now provides a practical "how to" guide to becoming one of those "ones." If you haven't bought it yet, do. If you have it, don't forget to pass it on when you've read it so that someone else can be a success. As Goethe wrote, "Are you in earnest? Seize this very minute."

Bruce

In the early 1980s, I saw a report on television that changed my life: footage showing overwhelming starvation and destitution—the reality for a huge portion of the world's population. As I was assaulted by images of small children with distended stomachs, I was going through confirmation classes to join the Evangelical Lutheran Church of America (ELCA). My class did a skit on salvation, involving people of various faiths and backgrounds discussing how they had lived their lives. The point of this skit was that salvation is based on how we *live* our lives, rather than on what we *say* we believe.

It's often pointed out that our sympathy for others is a function first of family, then community, then nation, and that by the time we get to the world community, it's much harder for us to experience empathy. But my pastor pointed out that Jesus' only comments on salvation appear in Matthew 25, and that Jesus presents salvation as a function of living our lives for those farthest from our family—those for whom our empathy will be least developed. In Jesus' time (before jet airplanes and the concept of a global community), that meant the traveler, the hungry, the homeless, the sick, and the imprisoned—basically those least a part of your family, those regarded by society as outcasts.

Confronted with this view of salvation, I started to ponder the question, "If I lived in the developing world, hungry and without access to clean water, basic health care, or much else, how would I want someone in the developed world to behave?" My tenth-grade history teacher, Dr. Barbara Schindler, was also asking this question. She talked about how governments address problems of domestic or international suffering and want. If people are starving or lacking water and health care, shouldn't it be the first priority of government in a Judeo-Christian nation to apply the Golden Rule in a macroeconomic way: to feed people, provide them with clean water, and treat the sick?

When I graduated from high school in 1987, I chose to attend Grinnell College, a small liberal arts college in the heart of the Iowa cornfields. At Grinnell, first-year students arrive a week before classes begin, in order to figure out where everything is and to learn about groups they can join. I immediately signed up for various progressive groups on campus. There wasn't an animal rights group, or a vegetarian or vegan group, but the leaders of the groups I joined were vegetarians—for human rights and environmental reasons. One of my new acquaintances suggested that I read *Diet for a Small Planet*, by Frances Moore Lappé.

Lappé's basic point is that raising animals for meat, dairy, or eggs is vastly inefficient. Farmed animals are essentially treated as food conversion machines: put a certain amount of soy, corn, or other feed into the animals, and get back some meat, eggs, or milk. Of course, by simply leading their lives, the animals burn off the majority of the calories fed to them, and some of their caloric intake goes into bones, blood, and other bits that are not edible. Lappé goes into much more detail, but the gist is this: if you eat meat, you devour exponentially more resources in the form of all the grains, corn, soy, and other foods fed to farmed animals than if you ate those crops directly.

The waste of eating dairy products, eggs, and meat—funneling crops through animals—while so many of the world's people starved, had a powerful impact on me. These were people like me who simply had been born in another part of the world. Lappé's arguments about the basic inefficiency and pollution involved in raising animals for food convinced me that I couldn't claim to be an environmentalist if I still ate animal products. The power of both of these concepts working together convinced me to go vegetarian immediately.

In 1987, homeless advocate Mitch Snyder came to Grinnell,

preaching about the hungry and homeless around the world. He told us we should drop out of college and really help people. Mitch's words weighed on me for a few years, and, after taking my junior year abroad at the London School of Economics, I did drop out of college to work in a shelter for homeless families and in a soup kitchen in Washington D.C., for six years as a part of the Catholic Worker movement.

While at the Catholic Worker, I was given the book *Christianity and the Rights of Animals* by Rev. Dr. Andrew Linzey, an Anglican priest and professor of theology at Oxford University. Linzey's essential message is about the nature of being an ethical human being in the world today. He frames his argument in a Christian context, but the argument is universal and applies to all of us: we should, where possible, make kind choices rather than cruel ones, and we should lead our lives with a goal of making the world more compassionate. When I finished the book, I began to talk to everyone I could find about the simple fact that we make a choice every time we sit down to eat: do we want to add to the level of violence, misery, and bloodshed in the world, or do we prefer to make compassionate, merciful choices?

We all know that animals have a capacity to feel pain, just as humans do; all animals have behaviors, wants, needs, desires, and so on. Anyone who has shared his or her life with a dog or a cat knows that animals are interesting and interested beings with personalities. It seems to me self-evident that, regardless of whether a person has a religious faith, people of integrity should make kind rather than cruel choices. People of integrity should choose not to support cruelty to animals, and thus should choose to eat a vegetarian diet.

I read more, had conversations with many people about suffering in the world, and took a few months to go back to school to complete my degree. A few dear friends, Betsy Swart and Aaron Gross, made a persuasive case that working for animal rights and vegetarianism would be where I could do the most good. That's how I came to PETA.

Not too long ago, my dad sent me a quotation that roughly states, "Find a job you love, and you'll never have to work a day in your life." My father is a Dean at Rutgers University, the state university of New Jersey, and is the personification of someone who does what he loves. My mother was an artist, following her passion for beauty and political influence through art (check out her work at ThistlewoodPress.com). They both instilled in me a belief that we should get joy out of what we're doing. I loved the work I was doing at the Catholic Worker, and I am even more satisfied and joyful with my "job" at PETA.

At PETA, I give talks, do interviews, organize demonstrations, write letters, work on websites, and help grassroots activists be as effective as possible. The latter is my greatest joy—working with advocates across the country to promote vegetarianism and animal rights in effective ways.

This is a book on the importance of being happy, and the fact that being happy is much more than a function of acquisition; it's a function of meaning and efficacy. A transformative experience that helped me recognize the need for this book came at a talk I was giving, when Dave Bemel, one of the most dedicated grassroots activists in the country, told me that my encouragement and accessibility when he was a new activist are the reason he's continued all these years. Dave tables at every fair or show that comes along, organizes an array of animal rights projects in his community, and promotes animal rights with vigor and eloquence. I said to my wife, "If I accomplish nothing else, this alone would be enough."

In asking myself what I could do that might create more people like Dave and my other grassroots activist heroes, it occurred to me that the basic pitfalls of grassroots activism aren't covered in an accessible way in other books. And since the person who's been most helpful to me in working out my sense of these topics is Matt Ball, I asked him to collaborate on this project with me.

We hope that our reflections help you in yours.

Matt

I'd like to be able to say that it all started back in nursery school, where, dressed like César Chávez, I organized a boycott of California grapes, in solidarity with the workers. However, I actually became an activist relatively late in the game. Growing up, I was a big fan of Carl Sagan, and I dreamed of exploring the universe, expanding the frontier of human knowledge and vision. I started my undergraduate work to become a rocket scientist, with the plan of working for NASA. I hoped to make a good salary and enjoy the American Dream—a big house, travel, good restaurants, etc.

But fate intervened on the first day of college, when I met my roommate, Fred McClintock. Fred was a big, strong guy who was not shy about explaining his vegetarianism or what hidden realities my eating meat supported. After a false start, I went vegetarian—I simply found the cruelties of meat production too severe to continue to rationalize away.

Shortly thereafter, I met Phil Murray (a fellow aerospace engineering student, currently co-owner of Pangea Vegan Products—veganstore.com), and then Jack Norris, who was Special Events Coordinator for the Animal Rights Community of Greater Cincinnati. Because we'd not yet discovered the core facts discussed in this book—namely, that ninety-nine percent of animals that die in the U.S. are butchered to be eaten; that modern agribusiness is by far the greatest cause of suffering in the U.S.—we followed the standard approach of the time: scattershot, "do something, do *anything*" activism. From writing letters to fur protests, from giving talks to participating in civil disobedience, Jack, Phil, and I pursued whatever came up, with no plan for maximizing the impact of our efforts and no vision for how the world could be fundamentally changed.

While a Department of Energy Global Change Fellow at the University of Illinois, I served as president of the group Students for Animal Rights. It was there I met Anne Green, and we married shortly thereafter. Meanwhile, on a trip out West, Jack passed a huge feedlot and was moved to send me a postcard, saying that we had to do something to express our outrage and bring more attention to the plight of farmed animals.

With ten other activists, Jack and I held a three-day Fast for Farm Animals in front of a Cincinnati slaughterhouse (three days being the amount of time farmed animals often go without food before slaughter). We held a large banner saying "Stop Eating Animals." On the last day of the fast, some of us left the slaughterhouse to hold the banner near the University of Cincinnati. After the fast, Jack and I formed Animal Liberation Action, which became Vegan Outreach, and the rest is history (veganoutreach.org/articles/history.html). Much of what we've learned since then forms the basis of the specific advice contained in this book.

As in life, everything else in this book is underlined by two fundamental, irreducible factors: pain and happiness. The majority of people I've met have never suffered so badly that they thought they were going to die, or been in so much pain that they wanted to die. Due to a chronic illness, though, I have been in those situations on a number of occasions. Because of this, I have a heightened sympathy for suffering. In the United States alone each year, millions upon millions of individuals actually suffer death after horrible, agony-filled lives. Yet even though our choices are the cause of these horrors, the plight of these individuals is not known—let alone considered—by nearly everyone in the U.S.

Of course, few of us want to know about great suffering; we want to be happy. That is basic human psychology. I've also found, however, that *true* happiness isn't to be found in ignorance, nor in following the crowd in pursuit of the American Dream. Rather, deep, lasting happiness is to be found in having a consistent, meaningful narrative guide our life. It's been my experience that for a narrative to provide lasting purpose and happiness, it has to be well-grounded and thoroughly examined. I hope that this book is successful at sharing my personal development of this narrative.

The Joy of a Meaningful Life

In *Possessing the Secret of Joy*, the writer Alice Walker suggests that the secret of a joyful life is working against violence and oppression. Our experience supports this—we believe that happiness comes from positive accomplishment in the world: i.e., from successfully striving to make the world kinder and more just. In this book, we present activism as the way to a meaningful life, and we offer some reflections on how we can be more effective in our efforts.

Where to begin? There is a great deal of suffering and injustice in the world, and much of it is gratuitous, graphic, and stomach-turning. Sadly, the vast majority of the world's injustices are too constant to allow significant attention. Furthermore, most of these atrocities are well hidden from everyday view. The human psyche is not inclined to dwell on unseen unpleasantness. Some say that reality is too much for us, that our brains simply can't process just how bad things really are for so many—much less deal with the moral implication that this suffering has on how we lead our lives. Others point out that we evolved to deal with only a small circle of people, not vast issues.

If we can't really process and understand global suffering, how can we be expected to seriously work for change in a methodical, long-term way? Indeed, most of us don't do anything. Among those who do decide to take action, most are drawn to the urgent, to the extreme, or to the anomaly—pursuing campaigns where immediate change can be seen.

The inclination to ignore what's "bad," or to focus only on the immediate and familiar, is understandable, given basic psychology. History has shown that arguing against or trying to deny human nature is a fool's pursuit. But that doesn't mean we're condemned to simply follow the crowd. We believe there's a better way—one based not on a new diet, job, car, relationship, political party, religion, meditation, or crystals, but rather on an honest evaluation of our basic nature. By assessing the world as it is, we hope to show that a firm grounding in reality can lead to a new path, one that's both sustainable and fulfilling for us as human beings. It's not simple; it can be argued that inertia is the strongest force in human nature. Ultimately, however, it's worth embracing the challenge to walk this new path.

In this book, we offer practical suggestions and insights for anyone who wants to choose a meaningful life by working to make the world a better place. Having been involved in various causes over the years, we've made many discoveries—and even more mistakes. On these pages, we've attempted to provide the sum of our experience, so that activists won't need to make the same frustrating missteps, but can instead go directly into effective activism.

We recognize that many of the suggestions in the book are counterintuitive, and some you may find challenging—we certainly did! We ask only that you give these ideas open-minded consideration and thought. For reasons we explain in detail, our focus is on advocating for animals. Subsequently, we devote the majority of this book to animal advocacy, as well as tips for being an effective activist for that cause. However, most of the suggestions and tips could be easily applied to advocacy for other issues; if you're active in some other cause to make the world more just, we salute you and hope you'll find some useful insights on these pages.

In Chapter One, we discuss the reasons we feel advocacy leads to meaning in life, and lay out our first principles and subsequent reasons for choosing vegetarian advocacy. In Chapter Two, we discuss some general tips for all activists. Chapter Three contains specific strategies for animal activists. These topics are related to questions, concerns, and scenarios we've been exposed to as animal activists. In Chapter Four, we discuss some of our favorite activities for changing the world. In Chapter Five, we discuss why we believe animal liberation is not only possible, but inevitable, and why we should all choose to be a part of this work. In the appendices, we've included a few of our favorite resources, as well as specific questions regarding the "whys" of vegetarianism, the question of "humane meat," and a theory of ethics.

The most important ideas in this book are:

1) Happiness does not come from an easy or materially "rich" life. It comes from a thoughtful, meaningful life dedicated to changing the world for the better.

2) Given our limited time and resources, as well as our inherent biases, we should make our choices based on reducing as much suffering as possible, remembering that when we choose to do one thing, we're choosing not to do another.

3) Simply making the "right" choices for ourselves isn't enough. By effectively interacting with other people, we can make the impact of our lives exponentially greater.

4) Our lives should be an advertisement for a purpose-driven life.

5) We should be professional and have positive, thoughtful *conversations* with people, rather than monologues at them.

6) We should *revel* in the freedom and opportunity we have, the ability to be a part of something bigger, something *fundamentally good*.

We find Robert F. Kennedy's speech at the University of Capetown, South Africa (June 6, 1966) to be of continuing relevance today:

Few will have the greatness to bend history itself, but each of us can work to change a small portion of events, and in the total of all those acts will be written the history of this generation. It is from numberless diverse acts of courage and belief that human history is shaped. Each time a [person] stands up for an ideal, or acts to improve the lot of others, or strikes out against injustice, [s]he sends forth a tiny ripple of hope, and crossing each other from a million different centers of energy and daring, those ripples build a current that can sweep down the mightiest walls of oppression and resistance. . . .

For the fortunate among us, there is the temptation to follow the easy and familiar paths of personal ambition and financial success so grandly spread before those who enjoy the privilege of education. But that is not the road history has marked out for us. Like it or not, we live in times of danger and uncertainty. But they are also more open to the creative energy of [people] than any other time in history. All of us will ultimately be judged and as the years pass we will surely judge ourselves, on the effort we have contributed to building a new world society and the extent to which our ideals and goals have shaped that effort.

The future does not belong to those who are content with today, apathetic toward common problems and their fellow [humans] alike, timid and fearful in the face of new ideas and bold projects. Rather it will belong to those who can blend vision, reason and courage in a personal commitment to the ideals and great enterprises of American Society.

Our future may lie beyond our vision, but it is not completely beyond our control. It is the shaping impulse of America that neither fate nor nature nor the irresistible tides of history, but the work of our own hands, matched to reason and principle, that will determine our destiny. There is pride in that, even arrogance, but there is also experience and truth. In any event, it is the only way we can live.

Peter Singer updates Kennedy's thinking in *How Are We To Live?*, where he calls us all to live examined lives. We'd go further to say that an honest examination reveals a life worth living is dedicated to building a better world. Singer explains:

> Anyone can become part of the critical mass that offers us a chance of improving the world before it is too late. You can rethink your goals and question what you are doing with your life. . . . Most important of all, you will know that you have not lived and died for nothing, because you will have become part of the great tradition of those who have responded to the amount of pain and suffering in the universe by trying to make the world a better place.

And, finally, to paraphrase Martin Luther King, Jr.:

> The arc of history is long
> And ragged
> And often unclear
> But ultimately
> It bends towards justice.

We can each choose to be a part of that progress.

We can be the ones who bend the arc, who change the world.

History is now—let's get started!

CHAPTER ONE

Choosing Meaningful Action

Is there a reason to care about how we live?

In his book *How Are We To Live?* (1995), Peter Singer describes how the collapse of religious traditions has left a void in many people's lives: "When Sartre realized that life has no meaning, it was a shocking contention. Now, it is simply the normal understanding." Most Americans wouldn't agree that life has no meaning, and the vast majority claim to believe in God and a higher purpose. Observing actual behavior, however, one wonders whether people are living lives that can be reconciled with their belief in meaning.

More than fifty years ago, Catholic Worker Movement founder Dorothy Day argued that we show our faith in a deeper meaning by living our lives in a way that speaks to our values. Like Gandhi and many other great spiritual teachers from a variety of faiths, she believed that our actions indicate both our values and whether we actually believe in a higher meaning.

It seems to us that "getting ahead"—accumulating material wealth and possessions for oneself and/or one's family—has become many people's *de facto* meaning of life, regardless of what they say they believe. For many others, life appears to be only about getting to the finish line, about filling up time with television, sports, and whatever else. For many (if not most) people, life has become a race to acquire more stuff and a fight against boredom.

But are those who live for possessions or simply to pass time really happy? Is this the best way to live? And more critically, how should we evaluate *our* lives? We hope you'll agree that answers to these questions are important enough to pursue with honesty and humility. Our struggle with these questions is the starting point of this book.

Evolution and Insatiability

The first step in our quest for a meaningful life is to break down assumptions and dig to the roots of our motivation, i.e., why do we make the choices we do? Sometimes our choices seem to be consciously considered; we might even make lists, "pro" on one side and "con" on the other. But there's always more going on—there are aspects of our nature as human beings that point us in one direction or another. With a general understanding of our evolutionary baggage, we can better understand why we often find ourselves pursuing material goods and simply passing time. From there, we may be able to make some rational assessments of whether these pursuits truly are the path to happiness.

Simply put, it's human nature to desire *more*, to strive for a greater share, regardless of what we already have. Over the millennia, those who were satisfied with what they had were erased from the gene pool by our unfulfilled ancestors. Individuals who pursued and obtained the most (e.g., food, partners, children, and other signs of "wealth") were the ones who prospered. The connection between "having" and the continuation of one's genes wasn't conscious, but was manifested in the individual's drives and desires for more, a discontent with the status quo, and envy of those with more. As Robert Wright summarizes in *The Moral Animal* (1995), "People weren't, of course, designed to be relentlessly happy in the ancestral environment; there, as here, anxiety was a chronic motivator, and happiness was the always pursued, often receding, goal."

These innate desires, built into our genes over the course of many millions of years, haven't disappeared; they remain a part of the human genetic makeup. Sadly, there really is no such thing as "enough," since our bodies are programmed with a view toward possible, unpredictable scarcity. Much of humanity has reached a point that our genetic programming could not have predicted—we can be reasonably sure we will survive and provide for future generations.

The fact that human programming to acquire doesn't have an "off" switch can be verified by even a cursory look at the world's millionaires and billionaires. In fact, it would appear that nothing satiates the drive for accumulating—there's always more to have, and there are always those who are better off with whom we feel we must compete. If you're Donald Trump, you can simply compete with yourself to acquire even more. Evolution has left us with a nature that *pursues* without end.

Americans are now about twice as rich as we were in the 1970s, and the Japanese are about six times richer than they were in the 1950s, but neither population is happier now, according to scientific studies (see, for example, imomus.livejournal.com/175376.html). Similarly, even lottery winners revert to their former baseline of happiness (Gilbert, 2007; Haidt, 2006). The phrase isn't "the *pursuit* of happiness" for nothing! Ultimately, the perhaps counterintuitive (and hard to accept) fact is that happiness simply isn't to be found in possessions or wealth.

Once we recognize our ancient, innate drives, we can more clearly and logically pursue what's really important—what it can and should mean to be human. Rational analysis reveals the pitfalls of our evolutionary heritage and can free us from desires that prevent us from achieving sustainable peace and happiness—drives that leave us striving toward, but never achieving, lasting happiness.

As rational beings, we can make decisions about how to live our lives based on logical and consistent derivations from first principles—concepts that we rationally agree are important, defensible, and fundamental—rather than evolutionary baggage, inherent prejudices, or current societal norms. In other words, we can choose to author our life's story, rather than following the narrative set by our genes.

The Meaning of Life: Working Toward Positive Consequences

> It is easy for us to criticize the prejudices of our grandfathers, from which our fathers freed themselves. It is more difficult to distance ourselves from our own views, so that we can dispassionately search for prejudices among the beliefs and values we hold.—Peter Singer, *Practical Ethics* (1993)

Throughout history, many have claimed to know the meaning of life. They set forth their particular philosophy, rules, and dogma. Looking at options today, we find dozens of competing approaches from philosophers, preachers, psychologists, etc., each convinced that theirs is the right way.

If we are to make rational choices as the basis of our life, we not only have to understand our genetic heritage, but we also have to recognize our cultural programming. If we want to be free of encumbrances that keep us from true fulfillment, it's important that we seek to understand what is fundamental, rather than accepting the dogma *du jour*, the beliefs of our parents, the preaching at the local church, current social views, the most recent best-seller. Just as reason shows us the biological baggage we've accumulated over evolutionary time scales, reason also shows us our cultural encumbrances.

Reason allows us to rise above all of this by showing us a larger perspective, revealing a "rule of the universe," where no one's interests count for more than anyone else's. Putting aside inherent prejudices leads to equal consideration of interests. Interestingly, this is summarized by "The Golden Rule," which can be seen as a core tenet of many ethical and religious traditions. "Love your neighbor as yourself," said Jesus. "What is hateful to you do not do to your neighbor," said Rabbi Hillel, the great Jewish teacher from just before Jesus' time. Confucius summed up his teaching in similar terms: "What you do not want done to yourself, do not do to others." The *Mahabharata*, the great Hindu epic, says: "Let no person do to another that which would be repugnant to her/himself."

Over the course of history, through endless human explanations, proclamations, and interpretations, many ethical systems have lost whatever connection they once had to fundamental principles like the Golden Rule. Most ethical systems have become a list of dos and (more often) don'ts. Those that have thrived have often come with large helpings of protection for the powerful and the status quo. It is an abdication of our rationality and humanity to accept the rules and laws of our day and our society as settled truths. We rightfully shudder at past persecutions of people like Galileo, Susan B. Anthony, Frederick Douglass, and Martin Luther King, Jr., all of whom questioned the dogma of the day. It would be naïve to think that today's society has all the right answers.

Where are we led when we pursue an objective perspective? Honestly and thoroughly considering a universal view shows that, if you dig far enough, virtually all actions can be traced to a desire for fulfillment or happiness and a need to avoid or alleviate suffering. In other words, when push comes to shove, thousands of years of philosophy can be summarized in nineteen words: Something is "good" if it leads to more happiness, and something is "bad" if it leads to more suffering. This is simplistic, of course. Not every situation lends itself to a clear analysis of consequences. Some things may seem intuitively "right" and actually be "right," even if the immediate consequences aren't obviously better than the consequences of different actions. Yet, despite the fact that some situations are difficult to analyze thoroughly, in general, focusing on the *consequences* of the actions is the most consistent way to maximize good outcomes.

Similarly, some things simply *seem* wrong, regardless of any consequential analysis. People come up with examples that seem to defy simple utilitarian analysis as well. But if we are to make rational, defensible decisions that are free of biases, utilitarianism offers a useful, straightforward, objective perspective that can help us avoid being distracted by personal or societal prejudices.

Given that pain—physical, emotional, or psychological—is generally the single greatest barrier to happiness, alleviating pain and suffering is a reasonable first priority for people who want to devote themselves to making the world a kinder place. We are in no way discounting the value of pleasure, but in the end we agree with Richard Ryder, who states in *Painism: A Modern Morality*, "At its extreme, pain is more powerful than pleasure can ever be. Pain overrules pleasure within the individual far more effectively than pleasure can dominate pain." In short, an objective ethics argues that we should make decisions based on what leads to the least amount of suffering.

Once we recognize that suffering is fundamentally bad, and thus eliminating suffering is the ultimate good, we can each dedicate our life to reducing as much suffering as possible. From these primary principles, we can give up the futile *pursuit* of happiness and, instead, live our lives beyond ourselves, for what is truly important. We can transcend our genetic and cultural programming and experience the full potential of our humanity and the richness possible in our existence. From a rational, universal starting point, we can choose to author our life's story, rather than following the narrative set for us by our genes and our culture. We can rise above the self-centered and immediate. We can be a part of something *greater*.

Striving for Morality: Our Influence More than Our Actions

The most important human endeavor is the striving for morality in our actions. Our inner balance and even our very existence depends on it. Only morality in our actions can give beauty and dignity to our lives.—Albert Einstein, letter dated 1950, quoted in Howard Eves' *Mathematical Circles Adieu* (1977)

It's logical that a fundamental ethic involves the reduction of suffering, but often ethics aren't thought of in practical, applied, consequential terms. Rather, "ethics" are thought of only as rules and regulations. It's because of human weakness and prejudice that so much philosophy and religion have become associated with commands and dictates. If we want to be free to live fully, to write our own narrative, we can reject the norms of today and the prejudices of our biology and recognize the fundamental wrong of suffering in the world. We don't need some formal structure to tell us right from wrong; we don't need a priest, rabbi, imam, or philosophy professor to tell us that it's bad to let someone starve to death, that genocide is wrong, that it's a tragedy when children are orphaned by AIDS.

More importantly, however, a complete ethic isn't just about minimizing the bad impact *we* have in the world. The needed change isn't simply *personal*. Although our decisions regarding what to eat and wear, what kind of car to drive, or for whom we should cast our vote are important, they're not as important as our influence on others. That is our *real* impact on the world.

Think of it this way: if we buy only vegan food, or we vote for the candidate of our choice, or we buy only coffee, clothing, etc., that's been produced fairly and responsibly, that's one unit of goodness in that area. If, however, we advocate for our position, each person we influence to do the same thing will *double* the good that our choices cause! Once we have ten people on board, our impact on the world is ten times greater than the choices that we, personally, have made *or that we will ever make!* In just *one* day, with just *one* interaction, we can do as much good for the world with our influence as we can do with our personal decisions and choices over the course of *our entire lives*.

To make the world better, we can—must—do much more than just make good, ethical choices ourselves. We can expose injustice, solicit kindness from others, and work for widespread change and the adoption of moral policies. Every person we meet is a potential major victory. Our power to change the world is much more than we imagine; our potential is mind-blowing!

We have no excuse for waiting. Living ethically—pursuing meaningful action toward a better world by alleviating and preventing suffering—doesn't require any consensus. If we were the one suffering—imprisoned unjustly, enslaved with no rights, exploited because of our race or species—we wouldn't want concerned, thoughtful people to put off taking action until the next election or until a large group endorsed our cause. We don't have to change the government to change the world. We don't have to start a group or organize a campaign. We can each act today and every day.

How Will We Focus Our Energy?

If we agree that the meaning of life is to make the world a better place by exposing and eliminating as much suffering as possible, then the most critical question of our lives is this: how do we do the most possible good in a world where suffering is so widespread?

Again, a basic understanding of human nature can show us potential prejudices and blind spots that might impede us from being optimally effective. Each of us has a bias of concern toward self-interest, the known, and the immediate. This applies to activists just as much as to the general population. Most people working for a better world concentrate on others who are most like them or who are closest to them geographically and/or biologically. It's almost too obvious to warrant mention, but most people working on gay rights issues are gay, on women's rights issues are women, on civil rights are African-American, on anti-Semitism are Jewish, etc.

These causes are important, but they're also issues of self-interest for many. Even with causes such as child abuse, cancer, domestic violence, and so on, leaders are often individuals with personal experience (e.g., when celebrities experience a disease, either personally or through a loved one, they often become spokespeople). Charities working within the U.S. get much more funding than those who do work overseas. Work on behalf of exploited or suffering human beings receives exponentially more funding and attention than work on behalf of non-human animals, and demonstrations for human rights attract more people and more moral outrage than demonstrations on behalf of animals.

Some people point to dogs and cats as an exception. In 2007, when investigators pulled sixty abused animals, dozens of animals' corpses, and truckloads of dog fighting paraphernalia from Atlanta Falcons quarterback Michael Vick's property, there were loud and vigorous demonstrations denouncing his cruelty to animals. At the same time, though, there were demonstrations supporting Vick, both on the field and in the community. Many commentators argued that the issue was not worthy of the concern and attention it was getting; others argued he shouldn't be suspended from playing football. Obviously, no one would have been pro-Vick if dead and battered human beings had been found on his property, or if the rape racks had been for humans, rather than dogs. Of course, the numbers protesting his actions were still a tiny fraction of the numbers that turn out for an anti-abortion or anti-war rally.

Some have expressed surprise or even envy at PETA's multi-million-dollar annual budget. This, too, shows the degree of our species bias—as if we're surprised an animal protection organization could take in such a "lofty" sum. Think about it: the largest animal rights organization in the world has a budget of some tens of millions of dollars per year to work against all of the combined injustices against the more than ten billion land animals who are killed annually in the United States. Planned Parenthood took in thirty times that much for work on women's health; Catholic Charities took well over a hundred times more to work on poverty issues. One human disease—cancer—gets thousands and thousands of times more money devoted to it than is contributed to every single issue related to animal rights. (For a ranking from 2001, see csmonitor.com/2001/1126/

csmimg/charitychart.pdf; see GuideStar.org for current budgets of other non-profit organizations.) Indeed, our entire government is focused on human needs, and spends billions each year subsidizing animal agriculture (see ucsusa.org/news/press_release/cafo-costs-report-0113.html).

Guiding Principles

An understanding of human nature, along with the recognition of the primacy of suffering, leads to two guiding principles that we've found useful in freeing our advocacy from prejudice.

First, to maximize the amount of good we can accomplish, we should strive to set aside personal biases as much as possible. We should challenge ourselves to approach advocacy through a straightforward analysis of the world as it is, motivated *solely* by a desire to alleviate suffering to the greatest extent possible. If the amount of suffering in the world weren't so vast, other considerations would be warranted (e.g., maximizing pleasure). But as long as so many are suffering so horribly, eliminating as much suffering as possible must be our primary motivating factor.

Second, it's vital we recognize that we all have limited resources and time. It's a simple fact that when we choose to do one thing, we're choosing not to do another—there's no way around it. Instead of choosing to "do something, do anything," we must challenge ourselves to pursue actions that will likely lead to the greatest reduction in suffering. There are a myriad of worthy pursuits, and of course we appreciate anyone working to make the world a kinder place. However, given the above principles, we challenge everyone—including ourselves—to constantly strive to maximize the efficacy of our actions.

Striking at the Root

There are a thousand hacking at the branches of evil to one who is striking at the root.—Henry David Thoreau, *Walden*

Perhaps you've heard the story of the person who finds babies floating down the river a few times per day, day after day after day—saving some, missing most of them. Every day, she waits by the river, knowing there'll be babies to save. Sure enough, every day she pulls some of the drowning babies out of the river, and she feels good about her efforts—saving lives, every day—even as she mourns the many who drown. Finally, one day she thinks, "Who on earth keeps tossing these babies into the river?"

She walks upstream, finds the person doing it, and stops him. In that moment, she's saved all of the babies who would have been tossed into the river in the future, and becomes free to dedicate herself to something else that would be helpful in the world. There's much triage work to be done in our society—there are many drowning babies, as it were. And obviously the work of saving them is good. But we're convinced that if we can stop people from tossing babies into the water in the first place, we'll be more effective.

In concrete terms, we choose not to focus our incredibly limited time and resources on individual animals, however valuable and rewarding that work is. Rather, we seek to challenge the very structures of oppression against animals, and to work to dismantle the system that says animals are commodities we can eat. To do this as effectively as possible, we must set priorities and, given our limited resources, make some difficult, rational choices.

Setting Priorities

Peter Singer asks us in "The Singer Solution to World Poverty" (utilitarian.net/singer/by/19990905.htm) to consider the case of a man who just bought a new car. He paid $50,000 for the car and doesn't have it insured yet. His car stalls on a set of railroad tracks, and, before he can push the car off, he sees a small girl also on the tracks, oblivious to an oncoming train. He has to choose between moving his car or saving the girl. Obviously, if he chose the car, all of us would hold him in moral contempt. Singer asks: what is the real difference between this scenario and buying the car in the first place, when you could buy a perfectly acceptable car for $20,000 or less, leaving $30,000 to dedicate to poverty relief, which would save far more than one child.

Similarly, consider the example of someone who has just bought an extra pair of two-hundred-dollar shoes. She sees a child drowning in the river. If the person chooses not to jump in for fear of destroying her shoes, again, all of us would find her morally reprehensible. Yet the same moral conclusion can be drawn when it comes to buying a pair of expensive shoes that aren't needed in the first place, rather than giving the money to charity.

When applying this to animals, the comparison becomes even more stark, since, for just a few coins, you can put an illustrated, detailed, documented booklet in someone's hands, show someone *Meet Your Meat* (meat.org) through online advertising, or show them a thirty-second vegetarian commercial. It takes so little to be the animals' voice, yet few of us even consider utilizing the power we have.

Even though U.S. society is composed mainly of professed Christians, most ignore Christ's words to the rich man: "Go, sell all that you have, and give to the poor" (Matthew 19). In an attempt to update the principle for our often selfish society, Singer makes the case that a reasonable standard for most of us would be to give away twenty percent of our income. Will that hurt, given that we've grown accustomed to our current level of income? For most of us, it will. At the very least, it will require an adjustment. But can we do it without actual physical harm coming to us? For most of us, yes, we can. Organizations dedicated to reducing as much suffering as possible can use that money to make the world better—far more so than whatever we might otherwise spend it on.

Singer sums up this concept in *How Are We To Live?* (1995), writing:

In a society in which the narrow pursuit of material self-interest is the norm, the shift to an ethical stance is more radical than many people realize. In comparison with the needs of people starving in Somalia, the desire to sample the wines of the leading French vineyards pales into insignificance. Judged against the suffering of immobilized rabbits having shampoos dripped into their eyes, a better shampoo becomes an unworthy goal. An ethical approach to life does not forbid having fun or enjoying food and wine, but it changes our sense of priorities. The effort and expense put into buying fashionable clothes, the endless search for more and more refined gastronomic pleasures, the astonishing additional expense that marks out the prestige car market in cars from the market in cars for people who just want a reliable means to getting from A to B, all these become disproportionate to people who can shift perspective long enough to take themselves, at least for a time, out of the spotlight. If a higher ethical consciousness spreads, it will utterly change the society in which we live.

We Can Do It

Take sides. Neutrality helps the oppressor, never the victim. Silence encourages the tormentor, never the tormented.—Elie Wiesel, Nobel acceptance speech (1986)

Consider this: the people we admire are not those who went along with the crowd, who did whatever was allowed by the norms of their times. Rather, the people we rightly respect are those who stood up to the prejudices of their society. Dr. Martin Luther King, Jr., Dorothy Day, Mohandas Gandhi, Susan B. Anthony, and so many other individuals changed their world. We are all called to do no less.

In the face of so much suffering, it can become easy to become despondent and to think that we can't change the world. But if we break our work into chunks, celebrate the "small" victories for what they really mean (e.g., turning one person vegetarian changes their entire life forever and makes a massive, positive impact in the world), and keep ourselves focused on our goals, we can realize what significant progress we're making. After decades of experience as activists, we're deeply and profoundly optimistic. Every day, we take inspiration in a review of the progress that has been won for social justice and animal protection (as we discuss at greater length in Chapter Five).

There are, of course, many potential targets for our activism: two billion people live without access to clean water; a billion don't have enough calories to sustain themselves; women in many parts of the world suffer unjust treatment and violence; our fellow creatures are abused and slaughtered.

These are a few of our society's current practices that, we're convinced, future generations will look back on with the same sense of incredulity we reserve for past atrocities like slavery and witch burnings. We are called to be like those we admire for standing up against the prejudices of their day.

Why Vegetarian Advocacy?

Because our singular goal is to have the greatest impact on the amount of suffering in the world, we've chosen to dedicate our lives to exposing the cruelties of factory farms and industrial slaughterhouses while promoting a vegetarian diet.

Emphasizing factory farms and dietary change is not our "personal issue." We have no special affinity to farmed animals over other animals (or human beings). Rather, this conscious choice follows directly from our fundamental guiding principles: 1) We want to maximize the reduction of suffering, and 2) We know that, by choosing to do one thing, we're choosing not to do other things.

Our experience has shown that promoting vegetarianism

offers the most effective and efficient way of decreasing overall suffering, for three basic reasons—the sheer number of animals, the enormous amount of suffering involved, and the opportunity the issue presents.

The Numbers

The number of animals raised and killed for food *each year* in the U.S. alone *vastly* exceeds any and all other forms of exploitation. The numbers are *far* greater than the total human population of the entire world: more than ten billion land animals are consumed in the U.S. each year, while the human global population is just over six billion. Approximately ninety-nine out of every hundred animals killed in the U.S. each year are slaughtered for human consumption. From a statistical standpoint, every animal killed in the U.S. dies to be eaten.

The Suffering

Of course, if these billions of animals lived happy, healthy lives and had quick and painless deaths, then our concern for suffering might lead us to focus our efforts elsewhere. (See our appendices on "Humane Meat" and "A Theory of Ethics" for more on the ethical questions surrounding killing animals to eat them.) But animals raised for food in the U.S. must endure unimaginable suffering. Indeed, perhaps the most difficult aspect of advocating on behalf of these animals is trying to describe the indescribable: the overcrowding and confinement, the stench, the racket, the extremes of heat and cold, the attacks and even cannibalism, the hunger and starvation, the illness, the mutilation, the drugging and breeding that create animals who can't even walk (e.g., "Farmed Chickens Can't Walk; Just Grow Them in Vats Already," blog.wired.com/wiredscience/2008/02/chickens-cant-w.html)—in short, the near-constant suffering and horror of every day of their lives. Indeed, every year, hundreds of millions of animals—*many times more* than the total number killed for fur, housed in shelters, and locked in laboratories combined—don't even make it to slaughter. They actually *suffer to death*.

The Opportunity

If there were nothing we could do about these animals' suffering—if it all happened in a distant land beyond our influence—then, again, our focus would be different. But anti-factory-farming/pro-vegetarianism advocacy is the most readily accessible option for making a better world. We don't have to overthrow a government. We don't have to forsake modern life. We don't have to win an election or convince Congress of the validity of our argument. We don't have to start a group or organize a campaign. Rather, every day, *every single person* makes decisions that affect the lives of these farmed animals. Helping people change leads to fewer animals suffering in factory farms. By choosing to expose the horrors of modern agribusiness and promote vegetarianism, every person we meet is a potential victory.

All thoughtful people want to see the world become more just and peaceful. Nearly everyone is worried about injustice and violence and wishes they could do something to stop it. What can we do about starvation and AIDS in sub-Saharan Africa? We can donate money, write letters, or try to get the government to intervene and give more aid. But all of those actions, though well-meaning, are far removed from having a measurable effect. On the other hand, three times each and every day, we make a concrete decision about who we are in the world. We each answer the question: "Do I want to add to the level of violence and misery and bloodshed in the world, or do I want to make a kind and compassionate choice?"

Simply put, the meat industry is violence we can either support or help stop. Every time we sit down to eat, we have the opportunity to have a profound impact on the world. Every meat-free meal is a blow against factory farms. Every time someone notices we don't eat meat, we're providing the animals a voice. It's hard to imagine any other choice we can make with such far-reaching effects.

It's very powerful to realize that following a vegetarian diet and setting a cruelty-free example allow us to take a stand against violence and suffering. *Every single time* we order from a menu, go shopping, or open up the refrigerator, we stand up for compassion. It is even more powerful to realize that every day, we can multiply that impact through our advocacy.

Seeing the Unseen

Our goal is to put aside our personal affinities, and instead focus solely on the suffering of others. In doing so, we've found that the above three points—the numbers, the suffering, and the opportunity—when taken together, are a logically compelling and, indeed, an irrefutable argument for working to end factory farming and promote vegetarianism.

Paul McCartney has pointed out that, "If slaughterhouses had glass walls, everyone would be a vegetarian." This concisely captures the main problem of vegetarian advocacy: people don't have to see the animals they eat being imprisoned in factory farms and butchered in industrial slaughterhouses. Someone can order a chicken sandwich, and to that person, it's just a sandwich. Even detailed, take-home illustrations, videos, and information about factory farms don't always stick with every individual. Society is set up not only to conceal the realities behind meat and divorce it from the actual animal, but to celebrate inanimate pieces of meat in and of themselves.

Similarly, if the realities of factory farms and slaughterhouses were as visible as the meat they produce, all thoughtful, compassionate individuals would be vegetarian advocates. Strip away the elaborate concealment and we'd see that nearly all animals exploited in this country suffer and die to be eaten.

With the animals visible, it would be apparent that every single person can make a direct and massive positive impact, simply by choosing to eat kind foods, rather than cruel foods.

Our inherent prejudice in favor of familiar animals (and for those whose suffering is immediately in front of us) has led those concerned with animals to spend hundreds of millions of dollars on cats and dogs each year, while focusing very little on the billions of domesticated animals slaughtered to be eaten. This is why one email plea about homeless pets in New Orleans is able to bring in much more money than any fundraiser for vegetarian promotion. Our hearts go out to familiar animals ("pets") who suffer through wars, natural disasters, and who are killed in shelters. We are, of course, glad that many people are compassionate toward these animals. Similar compassion and level of concern, exhibited efficiently and without prejudice, however, could have an exponentially greater impact.

For example, assuming only a very conservative one percent rate of change from the booklets that groups like Vegan Outreach and PETA distribute, there are tens of thousands of vegetarians in the world who would otherwise have eaten meat for the rest of their lives. Since the average American consumes about three dozen factory-farmed birds and mammals a year (and even more aquatic animals), distribution of these booklets has led to *many hundreds of millions* of birds and mammals being completely spared from the horrors of factory farms over the next fifty years. And that's assuming only a one percent total rate of influence and no multiplier effect (i.e., that each of these new vegetarians doesn't influence anyone else)!

In fact, there's reason to believe that the conversion rate is quite a bit higher, and that the multiplier effect is very powerful. PETA surveyed people who received their vegetarian starter guide, and responses indicated that more than eighty percent of non-vegans changed their diet, with twenty-three percent going from meat eater to an entirely vegetarian or vegan diet after reading the guide. Clearly there's some self-selection in survey responses, but these results indicate that the rate of change is probably greater than one percent, and that even those who don't immediately go entirely vegetarian may be cutting back on their meat consumption (please see animaladvocacybook.com for more details).

Because there are so few of us who look beyond the familiar and immediate, recognize the magnitude of the suffering caused by eating meat, and understand the opportunity vegetarian outreach presents, we have a special obligation to do the hard, intangible, and often unrewarding work of removing the walls that hide the atrocities of factory farms and industrial slaughterhouses. We need to be the vanguard, working as much as possible to abolish, totally and forever, the horrors of modern animal agriculture.

CHAPTER TWO

Effective Advocacy

When some people become aware of the vast cruelties of modern agribusiness, they often react with great passion. Some animal activists insist, "Animal liberation by any means necessary!" We applaud determination to take action, but recognize that millions of people before us have been outraged and furious with the state of the world. Since we joined the animal advocacy movement, most others who have become active have burned out and quit. Yet, we still meet people who say, "I'm willing to do anything!"

In this case, we need to ask ourselves these questions:

- Are we willing to give up—i.e., refocus—our anger?
- Are we willing to direct our passion, rather than have it rule us?
- Are we willing to put the needs of the suffering before our own desires?
- Are we willing to do the hard work of being effective, rather than just being active?
- Are we willing to do the necessary work, instead of spending a lot of time talking or trying to change other activists?
- Are we willing to accept slow change over no change?

In this chapter, we review these questions and discuss general strategies for effective advocacy. The comments can apply to activists for many causes. In the next chapter, we'll look more closely at some issues that apply specifically to animal advocacy.

Understanding Resistance and Being Open to Others

Hope for a better world starts when we learn how to get past our wall of denial, our desires, and our fear of change. Then, we must learn how to help others do the same. To succeed in freeing people to express their compassion—to open their hearts and minds—our interactions must be rooted in empathy and understanding: working with an individual's motivations, fears, desires, and shortcomings. Instead of approaching with a "fighting" mindset—which necessarily makes people defensive and closed to new ideas—we can approach others with information and options that they can digest in their own time and act upon at a sustainable pace.

The most effective advocates are, at heart, psychologists. Those who are most successful in fostering change understand that each of us is born with a certain intrinsic nature, raised to adopt certain beliefs and taught to hold specific prejudices. Over time, we each discover new "truths" and abandon others; we mix and match, supplement and refine, continually altering our collection of attitudes, principles, and values.

Even though we can recognize that our beliefs change over time, most of us are likely to believe that our current set of positions and opinions are "right," our convictions are well founded, our actions justified, and that we're basically "good." Even when, years later, we find ourselves reflecting with bemusement (or horror) on previously held beliefs, it seldom occurs to us that we may someday feel the same way about the attitudes we hold now.

It's interesting to read Mohandas Gandhi's autobiography, *My Experiment in Truth*, and to think about the man in his early years—wealthy, meat eating, living for himself. At that time in his life, it wouldn't have occurred to him that he would become the individual we now know as the Mahatma ("great soul"). More importantly, though, the young Gandhi certainly thought he was a good person. Similarly, all the great visionaries had early lives during which they didn't think, "I need to live differently." The same is possibly true for many of us; few of us reading this book can tell where we'll be in ten years, or how much impact we'll have in the world.

To have this impact, we recognize that real and lasting change comes from opening people's hearts and minds, allowing them the freedom to explore new ideas and new ways of viewing the world. Unfortunately, there's no magic way of doing this. The simplest way to encourage other people to open their hearts and minds is for us to open our own hearts and minds—and not just for the sake of advocacy or argument. Rather, we must be open and able to consider sincerely what is said during interactions with others. An open heart and mind is the best position for a serious advocate, because no one has all the answers.

The Importance of Optimism and Likeability

An important book for activists is Dale Carnegie's *How To Win Friends and Influence People* (1998). Although dated in its examples, the book's insights into human nature and interactions are very useful.

We're well aware that many advocates react strongly against the idea of reading a book with a title that seems so Machiavellian—winning and influencing. However, the book isn't about manipulation of others; it's more about basic good manners—things that can slip our minds when we're dealing with others who don't see the world in the same way we do. We owe it to those suffering to be as effective in our work as top salespeople are at making money.

Perhaps the most critical Carnegie principle for activists is that we should be optimistic, energetic, and positive in our demeanor. It's difficult not to be sad or angry about the amount of suffering in the world. But our question should not be, "What is justified?" or "What is an appropriate response to the facts?"— we can certainly justify being angry, negative, or depressed. Rather, the question we should ask in every situation is, "What will be most effective in helping those who are suffering?" For example, we ask, "How would a hen in a battery cage or a pig in a sow stall want me to behave?" Depression and anger, however understandable, won't be as effective at opening people's hearts and minds as a good-natured attitude. Think of the people who are the most influential—they're the ones who are smiling and upbeat.

When Bruce talks about these issues in presentations to activists, he puts up some images on the screen of Nelson Mandela and Martin Luther King, Jr. Both men are smiling broadly. Reading their works, their optimism is infectious. They knew they would win, and they inspired those around them. If we strive to be like them, we'll have the most impact.

In his book *The Tipping Point* (2000), Malcolm Gladwell analyzes the people who turn fads into trends. He has found that, in every case, they are friendly and optimistic. They express a genuine interest in others, which is returned by the people they interact with. Gladwell also offers many anecdotes showing that *how* things are said is even more critical than *what* is said. Whether they admit it or not, people are deeply influenced by body language and tone of voice. A positive tone and upbeat demeanor are far better at influencing people than being gruff, self-righteous, or angry.

Everyone has met an activist for whom malaise, anger, and/or misanthropy is worn as a badge of honor. Often, these activists have recently learned about the level of suffering, the awfulness from which many of them were shielded for so many years. Perhaps they feel it would be wrong to be happy when there's so much suffering. That's a fair enough response to the tragedies of the world, except that, again, a positive demeanor and happy attitude *will* make you a more effective advocate. Our effectiveness is the bottom line.

Often, the discussion of animal rights and vegetarianism online shows this anger. When pro-meat rationalizations are posted by meat eaters, often in angry and disparaging prose, the pro-animal forces respond in kind—by being belittling and judgmental. If the exact same points were instead made in a positive and affirming way, they wouldn't alienate readers, and might even win over adversaries.

Although we can't simply will ourselves to be happy, we can take steps in that direction. Research shows that simply by forcing ourselves to behave as though we feel a certain way, we can end up feeling that way (Layard, 2005). In other words, if you smile when you're actually sad, you send a signal to your brain that you're happy, with the consequence that you can actually become less miserable. To put this into practice, just before giving a talk, responding to a blog post, doing an interview, or going out leafleting, take time to smile into a mirror and laugh out loud. It sounds odd, but it can help put you into an upbeat frame of mind.

Strive for Balance

You should do what you can to ensure you're an activist for the long haul. If you sense you're burning out, get help. There's no honor in being depressed. There's nothing shameful in seeing a counselor. Yes, the state of much of the world is sad, but the proper response is to work to alleviate the suffering, not to withdraw and abandon the work that needs to be done.

Some people have asked how we can make jokes or be happy when there are so many suffering so terribly. But having a sense of humor is in any advocate's best interest, not only because it makes our example more appealing (and thus more effective), but also because it aids in avoiding burnout. The most successful activists we've known—those who survive and thrive over the long haul—are the ones who never lose touch with their sense of humor.

Another way of adding balance to our lives is getting involved in different organizations. Doing so also allows us to become a voice for change in these communities. Many activists are members of a religious community, a book group, a gym, or a basketball team. Great ways to start a conversation about your issue are to wear a T-shirt or button, have a cause-related bumper sticker on your car, and carry literature with you everywhere so that you always have something to hand to people if you end up in a discussion. Your workplace can also be a great place to talk with people about the issues. Just put a postcard or sticker on your wall or in your cubicle, and questions will follow!

We should revel in the many opportunities we have. If we allow ourselves to be miserable because of the suffering in the world, we add to that suffering. As long as there is conscious life on Earth, there will be some suffering. We can choose to add our own fury and misery to the rest, or we can work constructively to alleviate suffering as the basis of our joyous, meaningful, fulfilled lives. When something is upsetting—a personal problem has us losing sleep, some new injustice seems to defy comprehension, or even when we receive a nasty email from another activist about some perceived slight—we try to remember the words of the activist Emma Goldman: "Don't mourn! Organize!"

Effective Time Management

Given the state of the world, there's no shortage of issues that make demands on a thoughtful person's time. To be effective at making a difference, we must be effective in managing our time. If you're thinking, "I'm much too busy to learn about time management," then you're actually far too busy *not* to. Allowing our efforts to get wrapped up in the seemingly immediate and pressing, we're squandering opportunities and lessening our capacity to improve the world.

There are two critical lessons of time management we'd like to address here. The first, from Steven Covey's *The Seven Habits of Highly Effective People* (1990) is the importance of freeing ourselves from "the tyranny of the urgent." Basically, Covey suggests that most of us are so busy with the endless deluge of whatever comes up next—email on our screen, the phone ringing, and so on—that we don't have time to focus on accomplishing something substantial, something that really makes forward progress.

One example Covey offers resonates with most readers: if you're talking with someone at your desk and someone else comes by to see you, they wait or come back later. But most people, if the phone rings, pick it up. Why is that? The phone is seen by many as an interruption that must be accepted. Sadly, email is becoming like those phone calls, except there are far more emails than phone calls.

Covey breaks all activities into four quadrants (combinations of urgent and important), and challenges us to live most of our lives in quadrant two—activities that are not urgent but are important. It's also good to do things that are urgent and important, but if they're only urgent because you put them off, it's worth figuring out how to minimize future urgency in projects. The real trick is to minimize the time you spend on relatively "unimportant" tasks, especially urgent and unimportant tasks. One trick that works for us is to save up these tasks to do in chunks of time, so that you can complete them all at once.

The second lesson is to focus on the task at hand. On the television show *M*A*S*H*, the character Charles Emerson Winchester III sums up this concept perfectly when he explains, "I do one thing at a time. I do it very well. And then I move on." We have entered the age of multi-tasking, but for the vast majority of us, multi-tasking is counterproductive. Studies consistently show that if you do seven things in sequence, you will do them more effectively and more quickly than if you try to do more than one of them at a time. If you try to do two things at the same time, to do them both well will take longer than if you did them sequentially. Basically, the number of things we're working or attempting to concentrate on is inversely related to how well and quickly we can do them.

When Bruce speaks at conferences about these issues, he often asks people if they find working on a plane to be more productive than any other time. Pretty much everyone who works on planes has this experience—it's an enforced time to focus without interruptions. Of course, we could impose this sort of uninterrupted time on ourselves; we don't have to answer the phone, respond to email as it comes in, and so on. Learning this discipline, if we haven't already, will markedly improve efficiency.

All of this may seem obvious, but it's a sad fact that many of us who are working very hard aren't accomplishing nearly as much as we could if we were to implement some time management strategies. Basically, far too many of us simply do whatever is most immediate, rather than what will be most useful. Bruce speaks regularly about effective advocacy and time management, and every time he asks, "Do you ever have those very busy days, those days where you can barely take a break, and then you get home and your friend or partner asks, 'What did you do at work today?' and you honestly have no idea?" Every time, the entire room laughs knowingly. It's fine to have that sort of day on occasion, but each time you do, you've been less effective than you could have been.

When activism is your part-time endeavor (as it is for most activists), it's even more important to work as efficiently as possible. If we are to make real, substantial progress, we have to do better than fight every battle that arises. Turning off the onslaught of the "urgent" is often the only way to accomplish something substantial. If we don't know how our days are spent, that's a big problem; we're not focusing on critical tasks, and it's time to get serious about time management.

Here are a few more tips we've found effective:

- Keep *all* tasks to be accomplished on lists; commit no essential tasks to memory. Read *Getting Things Done* (2003) by David Allen for tips on how to do this most effectively.
- End each day with a list of things to accomplish the next day.
- Turn off email and don't answer the phone for hours

at a time in order to accomplish a larger or long-term project without distraction.

- Be flexible enough to make changes to your daily schedule, but not so flexible that you lose your control.
- Study time management; in addition to *Seven Habits* and *Getting Things Done*, the book *Never Check Email in the Morning* (2005) by Julie Morgenstern, and the video presentation to Google staff called "Inbox Zero" by Merlin Mann, are excellent (see animaladvocacybook.com for links).

Dress for Success

Some years back, an activist wrote in to *No Compromise*, a grassroots magazine about direct action, about how offended she was that other activists had asked her not to come to their demonstration against a poultry association because of her countercultural attire and look. Bruce replied in a subsequent issue, pointing out that although "looksism" is a form of prejudice, it's one where the consequences of changing our look (a fairly small compromise) can save animals from hideous abuse (a very positive effect). Bruce suggests that sporting a countercultural look at this type of event would only entrench the prejudices of the attendees and anyone watching on the evening news that those working for animals issues are not part of mainstream society. The suffering of those for whom we advocate outweighs any inconvenience we may experience in dressing a certain way.

Similarly, Paul Shapiro, the highly effective director of the powerful factory farming campaign for HSUS, recalls that as a young, angry activist, he said he would "do anything" for animal rights, but he wasn't willing to put on a suit and tie. Now, of course, he asks what the animals need, not what makes him *feel* most committed or radical.

In his excellent book, *Influence: The Psychology of Persuasion*

(2007), Dr. Robert Cialdini offers numerous examples of studies showing that how we look and carry ourselves is actually much more important to a successful encounter than anything we say in promoting our message. Cialdini discusses experiments where someone in a luxury car was able to sit far longer at a green light before they were honked at. Even people in work trucks and old Volkswagen Beetles (who many might think would be less impressed by wealth) allowed the nice car to sit longer before honking. Similarly, someone in a suit was more likely to be followed as they jaywalked than someone in jeans and a T-shirt, even by people in jeans and T-shirts! In surveys, almost *everyone* says they'd never behave that way; in fact, most people claim they'd honk at the luxury car first. But in practice, that's not what happens.

Similarly, in *Blink* (2005), Malcolm Gladwell discusses the speed with which most people form first impressions and make decisions about products and people. He argues, with abundant evidence, that appearance, tone of voice, and general demeanor are essential to one's likelihood of making an impact on a listener. People make early decisions about someone based on appearance and tone, and those decisions color the rest of the interaction. In tests, people preferred the taste of the exact same product on the basis of the packaging (i.e., changing packaging could raise or lower taste preference scores for identical products). Attractive packaging, for human beings as for products, is critical if we wish to influence people's habits and change their minds.

In the early 1990s, Bruce had a full beard and shoulder-length hair (see tinyurl.com/6b2zr2), and he only wore clothes that no one else would want. He was running a soup kitchen and homeless shelter, and frequently spoke in Catholic churches about peace and nonviolence. After re-evaluating his priorities and choosing a more mainstream appearance, he noticed that the quality of his conversations improved, and the respect of his listeners increased. Consequently, he became more effective an influencer of change.

Practically speaking, a $10 haircut and nice slacks and shirts (bonus points if you buy them at Goodwill) are significantly less expensive and time-consuming than tattoos, body piercings, and hair gel (an increasingly "conformist" uniform of a sort, when you think about it). All of us who are working for things that might be seen as going against the status quo should make sure our appearance doesn't detract from our message. Our message is usually difficult enough for people to accept, without us putting up any additional barriers.

The suffering we seek to end is *far* more pressing than always publicly advertising personal tastes and individuality. We are not the issue, and we should never distract from those suffering. There are times and places where green hair, body piercings, and ripped clothing are acceptable or even desirable. In most situations, however, when we reject mainstream society's standards, we're limiting our capacity to decrease the amount of suffering in the world. This is not a new concept; in *Rules for Radicals*, Saul Alinsky discusses the thousands of 1960s hippies who cut their hair, shaved their beards, and put on conservative clothes to "get clean for Gene"—Eugene McCarthy, the peace candidate for the Democratic nomination for the presidency in 1968.

Prepare and Practice

Fundamental to being an effective advocate is to prepare and practice having effective conversations with people. If you're uncomfortable talking with others, there are a variety of reasons for changing that, not the least of which is that those suffering can't speak for themselves. We're their advocates and we owe it to those suffering to have thoughtful and constructive replies ready for common questions.

There are only about a dozen things most people will say to justify their meat eating, for example. If you've been a vegetarian for more than a few months, you've heard them all. As the voice for animals on factory farms, it would be indefensible to be unprepared to discuss, for example, a question about whether it's natural to eat meat or where vegetarians get their protein. The difference between an effective and ineffective answer will be the difference between life and death for animals—each and every time you're asked that question.

There are many books and courses on how to become a successful communicator. Investing the time in one or more of these will give you a lifetime of improved advocacy skills. If you're currently in school, take some classes in communication, public speaking, and marketing. If you're not in school and feel that your communication skills could use a boost, join a Toastmasters class (they have them in just about every community in the country), or take a public speaking course at a community college. Any one of these options also gives you the additional benefit of a captive audience for your advocacy.

Some activists find it useful to listen to talks and recordings by other presenters when they drive. PETA President Ingrid Newkirk's keynote addresses from two animal rights conferences, her "nonviolence includes animals" presentation from Israel, and Bruce's "Veganism in a Nutshell" recording, are available free on PETA's website in the podcast section, and none is copyrighted. The more you listen to these arguments, the more you'll be prepared to make them yourself in conversation.

Another option is to have family and/or friends role-play with you: "Mom, I know you're not vegetarian, but I'm trying to become more effective in presenting the vegetarian case. Would you mind role-playing with me?" This has dual benefits and often works wonders, since you can make all the strongest arguments, such as asking how she feels about supporting cruelty so severe it would warrant felony charges if the animals in question were dogs or cats ("Does that argument impact you? Why not?"), etc. She's helping you, but she's also absorbing the information.

A useful (albeit embarrassing) exercise is to record this discussion with video. It's funny (and sometimes a bit mortifying) to watch yourself on video. Sometimes you'll discover you have a nervous habit, like touching your ear, and this will give you the chance to work on breaking it. Regardless, you can bet that once you're done, you'll be better able to make your case in any situation! If we believe in something strongly enough to have changed our lives, we must learn how to communicate our cause effectively.

Maintain Composure and Be Respectful

We should always attempt to maintain a non-defensive and non-antagonistic demeanor. All of us who've been a part of any advocacy community for long know activists who are less than kind (and are sometimes even downright bullying) in their interactions with others. We can understand this reaction, given the suffering in the world. However, if our goal is to create change, rather than express our personal anger, it's far more effective to meet everyone with compassion and engage them in a positive, rather than a negative, way.

Berating people has clear negative consequences: if we yell at someone who asks an aggravating question, we probably won't win that person over. This person is likely to tell others about the interaction, except that you'll be the aggressor in their story— the nasty, angry, activist. More importantly, we'll also alienate everyone within earshot.

Here is one example of the possible positive outcome on bystanders of behaving respectfully, even in the face of adversity: Matt was once tabling in a crowded lobby at a large agricultural university when a dairy farmer stopped by and started yelling. Matt calmly responded to the farmer, catching the attention of a passing Joe Espinosa. Joe listened, was impressed, subsequently joined Vegan Outreach, became a vegetarian then vegan, convinced a number of friends and family to become vegetarian, and has been a leading leafleter and activist for Vegan Outreach ever since.

Can you imagine if Matt had instead yelled at the dairy farmer and reacted defensively? Perhaps Joe wouldn't have even stopped. Perhaps he wouldn't have heard Matt's argument. As of this writing, Joe's handed booklets to more than 145,000 individuals. Think about the impact of this one instance of behaving respectfully in the face of disrespect: so many people have learned the truth of modern agribusiness. One instance of respect has made a tremendous difference in the world.

Of course, this same argument applies to letters to the editor and posts on blogs—we lose nothing by responding to vitriol with kindness, and we're far more likely to influence both the people we're emailing and the people reading. Beyond influencing an audience, we can try to empathize with the person who's being unkind to us. Some time back, Bruce received an email of truisms, and among them was this: "Be kind to unkind people (they probably need it the most)." Think about it: it's worse to be an unkind person than to be subjected to one. They are always miserable; you're only facing that anger and upset for that brief interaction with them.

If someone is clearly antagonistic, we can say, "I hear what you're saying," "I'm sorry you feel that way," or "Would you like to talk about that?" If we react in this manner, we'll be giving them a moment to embrace their better side. Even if most of these people won't change their view or lighten their antagonism, some will; even for those that don't, behaving in this way can only help. We know they won't come around if we act aggressively, defensively, or condescendingly.

Here are some more examples of this important aspect of activism. Long-time activist Anna Lesiecki tells this leafleting story:

> The most interesting interaction I had was when two guys came up to me and started yelling because there were leaflets on the ground between the entrance and

me, and that they liked meat and blah blah blah. I sweetly told them that periodically we walked around and picked up the littered leaflets (which really helps discourage others from tossing them on the ground), and that we hope people don't just throw them away, but rather that they take them, read them, and then choose to lead more compassionate lives. After maybe a minute or so of the rude guys yelling and me politely responding, one guy stopped and said, "Wow. You're really nice. We've been total jerks, and you've just been nice to us. So, tell us what your message is and why you're here." We then went on to have a perfectly civilized conversation, and they ended up taking a leaflet.

A high school teacher relates a story from class:

In my science class, a student asked a question about livestock and global warming. One question sort of led to another and we started talking about factory farming. At this point, one student—a nice kid who you wouldn't expect this from—started mocking me as I spoke. To almost everything I said, he replied, "But they taste soooo good!" and "You're making me hungry!" I kept my cool and smiled, though inside I was getting angry. But he came back a few days later telling me that he did some research on what we were talking about, and is going vegetarian. This just shows that sometimes the rudest people are those who feel the most uncomfortable when confronted by the facts.

No matter how right we are, we need to remember to ask ourselves: "What is most likely to have a positive impact?" When confronted with a new idea that calls into question how people have always lived, few are going to happily condemn their past and embrace change. Some will react antagonistically. Our role is to calmly give them the information, allow them to react to it however they will, and honestly and respectfully answer their questions.

Few people came to an enlightened view of the world by themselves and overnight. For many of today's leading advocates, it took a long time to change after their first exposure to the issues. For example, after being exposed to the realities of factory farms, it was more than a year before Matt became a vegetarian for good; he certainly wouldn't have changed if his initial reluctance had been met with disdain and mockery.

Matt's story is not unique. Not only does it show the downsides of anger and the benefits of patience, it also indicates that we shouldn't give up on our friends if they don't react to information exactly as we'd like them to. Shunning people because they don't immediately adopt our views not only cuts us off from the very people we need to reach, it also perpetuates the stereotype of the joyless fanatic.

CHAPTER THREE

Effective Advocacy for Animals

Career Questions

You don't have to work for PETA or Vegan Outreach or another animal advocacy group to make a positive difference. There are countless ways in which motivated individuals can realize their dreams, develop their talents, use their gifts, and be involved in effective advocacy. All but a handful of people in animal advocacy have careers that aren't directly related to their cause. In general, the movement needs talented and hard-working people in all areas, but perhaps the hardest positions to fill are support positions that command high salaries in the private sector (e.g., IT, web, finance, etc.).

Outside the realm of jobs with advocacy groups, one could go into marketing and purchasing to help advance the acceptance of new vegetarian meats; for example the "chicken" at Red Bamboo (redbamboo-nyc.com) is incredible. Why isn't it for sale in every grocery store in the country? Also, there might be a path in academic disciplines such as psychology or social and decision sciences that would be able to provide insights into advocacy. Finally, with China (see veganoutreach.org/chinese/) and India (see petaindia.com) becoming ever-larger consumers of meat, advocacy in those countries will be increasingly important.

Although we can't be certain of this trend, work on legal matters might become advantageous for the animals. Becoming a doctor can be a useful thing for advocates, since, in addition to the high-paying nature of many careers in medicine, a physician's opinion often carries more weight than a non-physician's. When Vegan Outreach co-founder Jack Norris went back to school to become a registered dietitian, his textbooks showed that alternative methods of research have replaced many methods using animals, and are more effective, safer, and less expensive. Many vegans shy away from getting into the sciences because they might have to do vivisection, dissection, or scientific procedures that currently use animal products. If animal activists get involved, though, they could lead successful and publicity-garnering challenges to the use of animals in education and help develop and improve alternative technologies.

Some people are moved to help animals and want to open an animal shelter or sanctuary; keep in mind, however, that there are many, many shelters and sanctuaries in the country, and they take huge amounts of time and money—time and resources that could be spent directly striking at the root. Even Gene Baur, head of the most successful farmed animal sanctuary in the country, notes this in his book *Farm Sanctuary* (2008): "For every Hilda, Hope, or Cinci Freedom, we are aware that another creature—in just as much pain and just as deserving of care—is being denied a place of mercy. There are always more animals than we can provide shelter for. Even if we had space for five thousand, or five hundred thousand, or five million, or a thousand times that number, it wouldn't be enough."

Many diverse and committed individuals have lent their talents to all aspects of PETA and Vegan Outreach. If it weren't for those who pursue other fields and financially support our efforts, however, none of our work would be possible. Because of our members' hard work in fields not directly related to animal rights, we have funds to work to reduce suffering. The more money we have, the more we can put into changing the world. Concerned activists are working against a multi-billion-dollar industry to change the hearts and minds of the public. We have the truth on our side, but we need funds in order to produce booklets and videos, orchestrate investigations, and advertise.

A major donor to Vegan Outreach, for example, decided in college to pursue a lucrative job in order to be able to fund advocacy work. This person's contributions have allowed the printing of many hundreds of thousands of booklets, which have been read by people who certainly wouldn't have been exposed to the realities of factory farms if not for this choice to pursue a high-paying job. By choosing that career, this financial activist has saved millions upon millions of animals—far more than he could have saved by passing out leaflets or (probably) any other job within an animal advocacy organization.

Striving to acquire great wealth can, of course, be the antithesis of a meaningful life. But those who earn large amounts of money have had an enormous impact on the amount of suffering in the world by contributing much-needed funds to organizations that work efficiently to expose and end exploitation and cruelty. Bruce has a dream that a billionaire (or multi-millionaire) will become an animal rights advocate and want to bankroll unprecedented outreach efforts. When that happens, society will be radically transformed within the course of a year or less.

Of course, if you have a job outside of advocacy, you can still be active. A number of activists have handed out thousands of vegetarian advocacy booklets, in addition to having a full-time job and a family. For example, Stewart Solomon, a full-time teacher and father, hands Vegan Outreach booklets to tens of thousands of students every school year. His efforts, and those of hundreds of other volunteers, have led to many new vegetarians, saving millions of animals from the horrible fate of factory farms.

Valuing Honesty and Good Information

Honesty is important in our information and approach. This can be difficult, because in today's society, it seems that if we don't scream the loudest, we're not heard. Because moderate voices are often drowned out, there can be pressure to exaggerate claims in order to advance our cause and to counter those who lie to defend the status quo. Furthermore, there's a natural tendency to accept any claim we want to believe without critical examination.

This can be especially true of vegetarians, who often feel isolated in a meat-eating world. In the long run, however, being uncritical in the information we believe and use for advocacy can harm our efforts, because we lose support from people who have come to realize that we're not objective, and we miss chances to convince people who are inherently skeptical.

More importantly, meat eaters are looking for some reason to dismiss new and potentially inconvenient truths—few people want to face the suffering in the world, let alone make real changes in their lives. If we offer them one morsel of questionable information among our sea of facts, they will seize on that to write off everything we say as suspect. It is imperative we present information the public can't disregard from sources they can't dismiss.

All advocates should strive to use only the best information. But getting accurate, complete, and unbiased information can be difficult, and we shouldn't simply accept something said by what appears to be a "reputable" source. For example, until 1999, some of the information in Vegan Outreach's pamphlets had been based on secondary sources that seemed to be fully documented. We assumed that they were accurate, but when we checked primary sources, the original reports often didn't correspond to what was being attributed to them by vegetarian advocates. Even first sources have problems—such as publication bias—and thus cannot be viewed in isolation.

There are several traps for advocates to avoid when it comes to choosing information to present. One of the most common we've encountered is the tendency of activists to start with a desired claim and selectively build an argument to support that claim. This can be particularly harmful when the claim is so at odds with conventional wisdom as to be easily dismissed, and thus anything else said is tainted or ignored.

Some advocates also extrapolate epidemiological data from another country to our own. Many activists use the results of research done in other cultures as though it necessarily applies to vegetarians in the U.S. A wide variety of confounding factors make many extrapolations from one culture to another difficult, including the fact that little of this research is done on actual vegetarians or vegans (see, for instance, veganoutreach.org/articles/healthargument.html). Fortunately, factual arguments, documented from third-party and industry sources, are plentiful. The arguments in favor of vegetarianism are extensive and ironclad. There's no need to exaggerate.

Animals, Health, or the Environment?

> Then vegetarians had a habit of talking of nothing but food and nothing but disease. I feel that that is the worst way of going about the business. I notice also that it is those persons who become vegetarians because they are suffering from some disease or other—that is, from the purely health point of view—it is those persons who largely fall back. I discovered that for remaining staunch to vegetarianism a man requires a moral basis.—Mahatma Gandhi, "The Moral Basis of Vegetarianism" (1931)

Sometimes, animal advocates suggest that, although *they* care about the suffering of animals, other people won't, and thus advocates must appeal to self-interest—health and the environment. Inevitably, the goal of these activists is to talk about something they feel will be more convincing to the general public.

This method of discussion and advocacy is problematic for several reasons. First, if our goal is to advocate for animals, we're unlikely to be effective as their advocates if we're not talking about what happens to them on factory farms and in slaughterhouses. Even if selfish arguments were honest and widely compelling *to the general public* (as opposed to seeming compelling to those who are already vegetarian), people adopting a vegetarian diet for purely environmental or health motives are less likely to help change the nature of the human relationship with other animals. Secondly, we recognize that one effective activist can do far more good than twenty (or even more) silent vegetarians who keep their views and information to themselves. Decades of activism have shown us that the cruelty argument is more likely to galvanize people to action than environmental and health arguments. By honestly making the animals' case, we serve to create more change, faster.

Yet many advocates feel that they have to focus on what is called "the health argument." Considering how many billions of dollars are spent every year on weight loss books and dietary guidelines, it's easy to think that the health argument for vegetarianism would be powerful and effective. While it's true that the "health food" industry is experiencing a boom, the effect has been to get people to eat an *additional* something (soy, red wine, dark chocolate), rather than to get them to *stop* eating something. Once you consider the evidence, what becomes very clear is that most people are looking for a "quick fix," instead of meaningfully changing their lives in order to improve their health.

A case in point: the leading health obsession in the U.S. has long been thinness, yet the percentage of Americans who are overweight continues to increase, and obesity-related diseases have skyrocketed in the past few decades. As reported in a 2006 U.S. Department of Health and Human services summary, from 1995 to 2005, "obesity prevalence increased significantly in all states." As the *Scientific American* reported, the percentage of adults who were overweight in 1980 was thirty-three; in 2005 it was over sixty. Those who were obese increased from fifteen percent in 1980 to nearly one in four in 2005. The number of states where obesity rates were higher than fifteen percent in 1991 was four, while in 2005, every single state in the union had numbers higher than fifteen percent, and the lowest rate was 17.4 (citations and links at veganoutreach.org/articles/healthargument.html).

It's important that, as we make the ethical case for vegetarianism, we offer an honest and thorough plan for staying healthy (see, for example, veganhealth.org). Historically, the nutritional information presented by animal advocates has led to many failed vegetarians—so much so that after two years of leafleting across the country, Vegan Outreach co-founder Jack Norris decided to go back to school to become a registered dietitian in order to honestly evaluate nutrition research firsthand, rather than promote the spin given by vegetarians.

Of course, a well-planned vegetarian diet does have a lot going for it! In many ways, it's easier to be healthy when consuming no animal products (easier to lose weight, lower cholesterol, etc.). However, simply advocating that people stop eating animal products is less effective than laying out a thorough case for optimal vegetarian health (maintaining appropriate levels of vitamin B_{12}, zinc, calcium, essential fatty acids, etc.). If we want to do our best to prevent suffering, it's good to have an understanding of vegetarian nutrition, and then present an honest, unbiased evaluation of the nutritional aspects of a vegetarian diet. Doing so leads people to trust that we're not partisan propagandists. Honest, thorough information also creates healthy examples of vegetarianism, thus leading to more effective spokespeople for the animals.

On the other hand, awareness of the impact of modern agribusiness has increased, especially recently, in the environmental movement. For example, United Nations scientists found that eating meat is "one of the top two or three most significant contributors to the most serious environmental problems, at every scale from local to global." Specifically, the 408-page report, "Livestock's Long Shadow," noted the meat industry's contribution to "problems of land degradation, climate change and air pollution, water shortage and water pollution, and loss of biodiversity."

A similar argument applies to global poverty and the food crisis. The United Nations Special Rapporteur on food policy called the diversion of crops to be turned into biofuels "a crime against humanity," based on a hundred million tons of corn and other crops that could feed people instead feeding our cars. But more than 750 million tons of cereal crops are diverted from the mouths of the hungry into the mouths of chickens, pigs, and other animals—and that doesn't even include the eighty percent of the global soy crop that is also fed to farmed animals. As the Worldwatch Institute puts it, "[M]eat consumption is an inefficient use of grain—the grain is used more efficiently when consumed by humans. Continued growth in meat output is dependent on feeding grain to animals, creating competition for grain between affluent meat eaters and the world's poor" (please see animaladvocacybook.com for more details). As we strive to solve problems ranging from global warming to global hunger, more people will be forced to confront the realities of meat production. Although often reluctantly and slowly, more organizations and individuals may come to the inevitable conclusion that the developed world's subsidies for the meat industry and consumption of vast quantities of meat should be discouraged.

That said, it's important to recognize both the impact of who makes these types of cases and how these statistics are interpreted by the general public. It is different when the United Nations issues a report than when PETA or Vegan Outreach tosses around claims that are clearly part of a pro-animal agenda. Perhaps more importantly, although environmental and resource statistics may seem, to us, to be clear arguments for strict vegetarianism, many people interpret them as pointing out the greater efficiency and environmental friendliness of chicken vs. beef (see links to examples of "eat more chicken" arguments at veganoutreach.org/gwthoughts.html).

Unless it is something easy and painless—like a small donation to tsunami victims, recycling a can instead of throwing it out, or eating more of a certain type of food—the majority of people are not inclined to change. Most people are extremely defensive and capable of great rationalizations when it comes to personal culpability. Human nature leads us to focus on problems caused by others and dealt with by others: "Yes, it's terrible how that company hurts the environment!" "The government needs to stop that abuse!" For example, no one ever turns away from a graph of relative water usage or waste production, but many do turn away from, say, a screening of the video *Meet Your Meat*. That isn't because the latter is the "wrong" message. It's because, unlike statistics of water usage or cancer rates, images of obvious cruelty cannot be ignored and forgotten; they demand real, personal change.

If you were to ask the average person on the street what their main concern is, factory farming wouldn't top the list. As advocates, however, we're not trying to reinforce people's existing concerns and prejudices. Rather, our goal is to show them the hidden truths, expanding their circle of consideration beyond themselves. The most common feedback Vegan Outreach activists receive from our booklets or video screenings is, "I didn't know what went on in factory farms."

Whole Foods CEO John Mackey explains his veganism this way: "Eating animal products causes other beings to suffer, and we don't need to eat animal products to survive. In fact, we're better off without them. I don't want to cause unnecessary suffering." It's a simple, powerful, and unassailable argument. Nothing that anyone says disproves this fundamental message, "Eating meat causes animals to suffer unnecessarily." Whether it's "The Bible says we can do it" or "What about abortion?" or "What about plants?" nothing counters the simple, undeniable fact that eating animals causes unnecessary suffering. We have found it to be the strongest argument for vegetarianism.

Applying the Socratic Method in Our Discussions

The suffering of animals on factory farms and in slaughterhouses is compelling to us. Consequently, we often want to do anything to cajole others into sharing our horror and outrage. However, after decades of advocacy, we're now convinced that this isn't the most effective way to convince people to change their behavior.

Remember the adage "Hear and forget, see and remember, do and understand"? As advocates, we want to do our best to move people to that third stage. We want to avoid lecturing others, since that turns them into passive recipients of information they can easily ignore. Instead, we want to focus on having a conversation where our companion's thought process leads them to their own conclusion. By asking questions, we can help people to understand that making compassionate choices is a simple extension of the values they already hold.

For example, everyone opposes cruelty to animals (well, ninety-seven percent of Americans do, according to a 2008 Gallup poll—you do have to wonder about that other three percent). Convincing someone to go vegetarian should be simple, since eating meat supports horrible cruelty and isn't necessary. But if it were simple, everyone would already be convinced. Our efforts must be more effective. Rather than launching into a monologue about cruelty, however, we must lead people to recognize that what they already believe (cruelty is wrong) necessitates a change in diet.

When someone says, "Plants feel pain!" or "Animals eat other animals!" there are certainly many possible responses that would ridicule their argument. But no one who says these things is trying to be stupid; they often believe they've caught us in a moral inconsistency, generally because they haven't spent much time thinking about the issue. Because we *have* spent the time thinking about it, we must respond sincerely, even if we find the remark stupid.

A wonderful way to begin an answer to a "stupid" question is to reply, "That's a good question, but do you think that . . ." or "I used to ask that same question, but then I realized. . . ." This sort of segue validates the other person, shows anyone else listening to the conversation that we're respectful, and continues the discussion in a way that will be far more effective than any other method.

We know there are many situations when we haven't said a thing and people are still on the defensive. Deep down, many people know animals suffer, and that it's wrong to eat our fellow creatures. Thus, they feel inherently judged and guilty when they find out we're vegetarians. Anger is, in fact, a reasonable response to the realization that something you've been doing your entire life is morally indefensible. Therefore, we welcome their reactions; we would much rather have someone angry than apathetic about what we're saying. We can't let their defensiveness or personal attacks make us antagonistic—their anger is understandable and worth building on; an irate response from us would be counterproductive. If someone makes a rude remark, try to laugh and say, "Hey, you brought it up. Maybe you'd be willing to take a look at this booklet. I'd be happy to discuss this with you after you've had a chance to look it over."

Still, given the massive cruelties of factory farms and industrial slaughterhouses, it may seem imperative that we take every opportunity to make the animals' case as forcefully as possible. In the past, if someone asked, "Why are you a vegetarian?" we would immediately launch into a breathless diatribe:

Animals on factory farms are treated like machines. Within days of birth, for example, chickens have a chunk of their beaks seared off with a hot blade. Male cows and pigs are castrated without painkillers. All of these animals spend their brief lives in overcrowded spaces, breathing ammonia-filled air, many of them so cramped that they can't even turn around or spread a wing. Many do not get a breath of fresh air until they are prodded and crammed onto trucks for a nightmarish ride to the slaughterhouse, often through weather extremes and always without food or water. The animals are hung upside down and their throats are sliced open, often while they're fully conscious. I believe that if you saw how animals are suffering on factory farms and in slaughterhouses, you'd be horrified and you wouldn't want to support it. Plus,

you'll be healthier—you'll have more energy, need less
sleep, and just feel better if you're eating a vegan diet. . . .

And so on, trying not to take a breath for fear the other person
would use that opportunity to beat a hasty retreat.

There have been people who have adopted a vegetarian
diet after hearing this monologue; indeed, the arguments are so
strong that we're likely to convince at least a few people with
them, regardless of how we go about it.

But we can do better.

It can be hard to understand, and even harder to put into
practice, but when someone asks a question, the best tactic is to
ask them a question back, to listen to their answer, and to *have a
conversation* with them about the issues. Rather than "questions
and answers," what we want is to lead the conversation to where
our companion is answering their own questions. It's called
"the Socratic method." Socrates was killed because his method
proved so successful in influencing the youth of Athens—it's *that*
effective in helping people to think and change their views.

Here are some examples of answering with a question.

> *Question*: Where do you get your protein?
> *Answer*: There are some great vegetarian meats
>   on the market. Maybe I can recommend
>   something you'd like to try. What's your
>   favorite source of protein?
> *Question*: Why are you a vegetarian?
> *Answer*: Thanks for asking—I'm opposed to cruelty
>   to animals and don't like what happens on
>   modern farms. Do you eat meat?
> *Reply*: Yes, of course!
> *Answer*: Really, why?

The latter is Bruce's favorite. Bruce owns twelve "Ask me why
I'm vegetarian" t-shirts (it's the only t-shirt he wears), so he gets

the question a lot. He always turns the tables on the questioner, asking them if they eat meat and, if so, why. Most people have never even thought about it. Leading them to think about what it means to choose to eat animals is a very effective way of helping them challenge themselves on the issue.

As we've seen, almost everyone agrees that cruelty to animals is a bad thing; therefore, we like to focus the discussion on the fact that eating meat supports cruelty. We do this by allowing the person we're speaking with to recognize the importance of *their own position*, rather than just telling them about it. There are at least three reasons why responding with a question is effective.

1) People have to think about why they eat what they eat.

Instead of launching into a monologue, we give a brief explanation for why we are vegetarians (e.g., "I don't want to pay others to do things to animals that I wouldn't want to do myself"). Often, we then try to move quickly into a question, "Wouldn't you agree that mutilating an animal unnecessarily is wrong?"

There are a number of questions we can ask, such as, "Would you want to *watch* any aspect of what is required to get chickens, fish, pigs, cattle, or dairy and eggs to the table?" (You can offer an illustrated booklet if they claim not to know what goes on.) "Do you think it's OK to support that? Would you want to spend even five minutes in a slaughterhouse, with all the blood and horror? No? I agree, and that's why I've decided it's not OK to pay people to do things I couldn't even watch. Can you understand why I feel that way?"

These are all valid questions to ask others. Instead of telling people how immoral meat is, raising these questions with people in a conversational way can allow *them* to decide that eating meat is ethically unjustifiable. In short, they can come to realize that by eating meat, they're paying other people to do things none of us believe in or would otherwise support.

Most people will express remorse at the horrible things done to farmed animals on factory farms and at slaughterhouses.

Once we have the person agreeing to this, they're in a position to recognize the power of their own choices, "So why do you think people eat meat when they know they're supporting things like tiny cages, mutilations without pain relief, slaughter without stunning?" Then, wait for an answer. It can be uncomfortable, but fight the urge to fill in the silence. There's real power in having people answer this question.

2) People feel heard, so they return the favor by listening.

Dale Carnegie points out that few things are sweeter to someone than the sound of their own voice. People *love* to hear themselves talk, and they generally have a high regard for their own thoughtfulness. The converse is also true: lecturing people, instead of listening to and engaging them, causes them to be less interested in what we have to say, for the simple reason that they will be less likely to respect us and enjoy our conversation.

3) We'll be more effective if we know what we're up against.

For example, if we ask, "Do you oppose cruelty to animals?" how we proceed will vary radically if the person we're speaking with replies with "I give money to my local SPCA" as opposed to "I don't really see it as a primary concern." We need to know who we're talking with if we want to be as effective as possible.

The following is a real conversation Bruce had on a plane, the day he was working on this chapter.

Jane Doe:	Okay, I'll bite, why are you a vegetarian?
Bruce:	Thanks for asking. Can I ask you first, do you eat meat?
Jane:	Yes, not as much as I used to, but yes.
Bruce:	Really, why?
Jane:	Well, I like the taste.
Bruce:	That's interesting; does it give you any moral qualms at all to eat an animal?

Jane:	Well, I've never really thought about it. Why would it?
Bruce:	Well, for me, I don't want to pay other people to slit animals' throats open. Really, I don't want to pay someone else to do something I wouldn't want to do or even watch. Does that make sense to you?
Jane:	Well, I guess so, but don't animals eat other animals?
Bruce:	Sure they do, but doesn't that prove my point? Animals are doing what's natural to them, but if we're eating meat, we're supporting the cruelty of factory farms and modern massive slaughterhouses. Did you know that there are no laws to protect farmed animals from abuse, and that what happens to them would warrant felony cruelty charges if these were dogs or cats?
Jane:	Really, no laws? I hadn't thought of that. But where do you get your protein?

And so the conversation goes; Bruce is able to make all his points in a way that allows the other person to truly consider everything being said. *Nothing* is more important in our interactions with others than actual conversations instead of lectures.

In another situation, Bruce pulled out his laptop at a restaurant to get a phone number. He has a PETA bumper sticker on the back of his computer ("Free Vegetarian Starter Kit: 1-888-VEG-FOOD"). Four teenage boys approached him, asking, "OK, what's a vegetarian starter kit?" They seemed simultaneously shy and haughty. Bruce explained the vegetarian information and asked them if they knew any vegetarians, to which they replied, "Well, we're all hunters; we could never go vegetarian." Instead of attacking them for hunting, Bruce asked, "What do you think of cruelty to animals?" Naturally, they tripped over one another to condemn it. Bruce then had a good conversation with them about what happens to chickens, turkeys, and pigs on factory farms. They were horrified, and all four pledged never to eat birds again. They returned to their group with their vegetarian kits (Bruce had a stack in his bag—as all activists should), chatting about how they were going to explain their new dietary commitment to their parents.

Could this conversation have happened if Bruce had attacked them for supporting cruelty by hunting? Could it have happened if he'd not first got them to go on record in opposition to cruelty to animals? Probably not. In fact, it's likely it wouldn't have happened if Bruce had asked them to go completely vegetarian. (For the same reason, many activists leaflet with *Compassionate Choices* or *Even If You Like Meat* booklets, rather than *Why Vegan?*). Being vegetarian seemed so foreign to the teenagers at that moment, but since Bruce talked about how badly birds are treated (and many, *many* more birds are killed and eaten each year than mammals), they ended up agreeing not to eat them.

Every year, the average American consumes about one-tenth of a cow, one-third of a pig, one turkey, thirty-five chickens, and about fifty aquatic animals (mostly shellfish). She or he is also responsible for the output of one laying hen and one-thirtieth of a dairy cow. Based on the raw numbers alone, the best incremental step a meat eater can take for animals is to stop eating birds. And that's how we talk with people: we focus on cruelty to birds first. Once they've seen that they can make a step, it's much easier for them to move on to stop eating pigs, fish (especially farmed fish), eggs, cattle, and then dairy.

Few people adopt a vegetarian diet overnight. If we help more people change by accepting incremental evolution—preferably by no longer eating birds and fish first, then pigs, then cattle—we can help spare many animals tremendous suffering. Since most people will otherwise go about this the other way (giving up cows and pigs first) we do a real service to animals by focusing on cruelty to factory farmed birds first.

In the end, the case is straightforward. As mentioned above, according to surveys, ninety-seven percent of Americans want to see animals protected from abuse, and a May 2008 Gallup poll found that fully twenty-five percent felt that animals deserve "the exact same rights as people to be free from harm and exploitation" (tinyurl.com/5pyvco). Of course, people are generally thinking about dogs and cats rather than farmed animals, but it shows a strong base of ethical consideration. Since so many are compassionate to the animals they know, it's our job to be the ambassador for the animals they don't. If we can help people open their hearts and minds and show them our common opposition to cruelty, they can realize that agreeing with us should mean they adopt a vegetarian diet.

There is hope for a real change in the world if we learn how to help people get past their wall of denial and manifest their latent compassion. To succeed, our interactions with others must be rooted in empathy and understanding, working with and from a person's motivations, fears, desires, and shortcomings.

Maintain Focus

Animals deserve our single-minded attention to their plight, and we don't want to alienate possible vegetarians by arguing with them about their faith, politics, etc. Often when we're making the animals' case, the other person will try to do anything to get us onto some other topic, like abortion, religion, or some hypothetical situation about a deserted island. We must refine our responses to these distractions such that they lead the discussion back to unnecessary cruelty to animals.

However strong our opinions on other issues might be, we owe it to animals to focus the discussion solely on their suffering. If we make vegetarianism and animal rights a package deal that includes other issues, it will be easier for others to dismiss us. People who have radically different opinions about religion, foreign policy, or the death penalty can all agree that animals are not inanimate objects, can feel pain, and don't deserve to be made to suffer. The message of compassion and justice can appeal to everyone, and it's important to focus on that common ground in all our interactions.

Some animal advocates believe that people who don't share their political beliefs won't care about cruelty to animals. Consider, though, that Senator Bob Smith, one of the most conservative individuals to serve in the Senate over the past fifteen years, garnered unequivocal support from Humane USA, the political action committee set up by Humane Society of the United States. Senator Smith was the sponsor of many bills to help animals and spoke eloquently from the Senate chamber about human responsibility toward other animals. Republican Bob Dole was also much better on animal issues than Democrat Bill Clinton, whose first secretary of agriculture resigned after taking gifts from Arkansas-based Tyson Foods (which kills more than two billion chickens every year). While a representative, Clinton's second secretary of agriculture took more money from agribusiness than anyone else in the House. More recently, Matthew Scully, former speechwriter for President George W. Bush, left the administration to promote his book *Dominion: The Power of Men, the Suffering of Animals, and the Call to Mercy*, which eloquently argues the case for veganism and animal rights on conservative grounds.

One can quickly name half a dozen prominent conservative pundits and journalists (such as Pat Buchanan, George Will, Fred Barnes, Tony Blankley, Rich Lowry, Charles Colson, and G. Gordon Liddy) who've made strong statements on behalf of animal welfare, even farmed animal welfare. Our point isn't that we should all become conservative Republicans. Rather, our point is that liberal Democrats and conservative Republicans— and everyone in between—can agree on the issue of cruelty to animals.

Personal Purity vs. Effectiveness for the Animals

Most people deeply concerned with ending the cruelties of factory farms and industrial slaughterhouses are vegetarian: that is, they don't consume obvious animal flesh (pigs, sheep, birds, cows, fish, etc.). It's hard to imagine how an individual could claim to want an end to the unnecessary cruelty to animals inherent in viewing them as merely flesh, but still support raising and slaughtering animals for food. However, some activists take this concept and extend it to create a list of ingredients that define vegan.

When Matt first became involved in animal rights, the main concern in the community was often not with making a difference, but with making a list. Incredible amounts of time were spent determining if something was "vegan"—containing no hint of animal product at any level. The way to tell was to compare all of the ingredients on every product against lists of all possible animal products. This list eventually became an entire book, *Animal Ingredients A to Z* (1997), which was the best-selling book at vegan.com for years! Among many people (especially on early Internet talk groups), the main discussion wasn't about helping animals, but showing just how far an individual would go to be "vegan"—an endless loop of one-upmanship.

This way of defining veganism neglects several key points, not the least of which is that everything involves some suffering. At some level, anything we consume harms some animals. Almost all organic foods use farmed animal manure as fertilizer, adding financial support to animal agriculture. Planting, harvesting, and transporting any foods kill and displace animals. Bike tires and even some "vegan" shoes contain a small amount of a product derived from animals. Some sugar is processed with bone char, as is some water, and so on.

Some anti-animal activists will demand that animal activists and vegans account for animals caught up in combines or insects killed by pesticides, as though these deaths mean that vegans are no different than meat eaters. This makes sense if you believe "vegan" equals "pure." But if you see veganism as a tool for reducing suffering, the attempt at equivalence fails; it takes exponentially more pesticides and causes more deaths (planting, harvesting, transport, etc.) to grow food to funnel through animals than to eat the plants themselves.

Unless you believe being vegan is about being personally blameless, the entire premise—that if you can't be perfect, you should do nothing—is absurd. We could go out into the woods and live on nuts and berries as "level five vegans" (a concept from *The Simpsons* television show) or truly "do no harm" by committing suicide and letting our bodies decompose in a forest. Ultimately, though, these options would be far less effective at eliminating suffering than living in ways that could influence others to change their diets.

The goal can't be to totally eliminate suffering; the goal must be to make choices that cause the least harm. Obviously, this will mean never consuming the products for which animals are raised: meat, eggs, and dairy. But if whey or "natural butter flavor" is in the "less than two percent" category on an ingredient list, the connection to cruelty, while perhaps uncomfortable and aesthetically questionable, is negligible and is probably no more, calorie for calorie, than many "vegan" foods.

If you adopt an ethical diet, you no longer support the torment and slaughter of dozens of land animals every single year. Helping *just one more person* change their diet will save twice as many animals. If we believe our choices are important, being the most effective advocate for the animals is, by definition, *even more important*. The impact of our individual choices—several hundred land animals over the course of a lifetime—*pales* in comparison to what we have the potential to accomplish with our example!

But the reverse is also true: if we do something that dissuades another person from adopting an ethical diet—if, for example, our actions put up a barrier where we might have built a bridge—we hurt animals.

Currently, the vast majority of people in our society have no problem eating the actual leg of a chicken. It's not surprising that many people dismiss vegans as unreasonable, irrational, and self-centered when their experience includes meeting or knowing people who won't eat veggie burgers cooked on the same grill with meat, who scrutinize menus and interrogate waiters, discuss the origins of ingredients they've never heard of, and so on.

For example, if we're at a holiday party with meat eaters and we're talking about how we can't eat the bread because we don't know what's in it, or if we're at a restaurant and there's a veggie burger on the menu but we give the server the third degree about the ingredients or about how it was cooked, we're most likely doing more harm than good. By coming across as fanatical, we're turning people away rather than turning them on to the joys of compassionate living. We've just made eating ethically seem incredibly difficult, and have created barriers to the others at the table who might have otherwise considered the plight of animals. In this situation, others are unlikely to ask about our diet, and they're even less likely to consider it for themselves.

If you're worried about what you're going to eat in a restaurant, call ahead and figure out what meets your standards, and then order it with gusto. If you're worried about what you're going to eat at the office party, get on the catering committee or just bring along some great vegetarian food. Never make it seem like being concerned about animal suffering is a chore, an encumbrance, a barrier—because it doesn't have to be, and it shouldn't be. It's a wonderful statement about the power of our choices.

As this is a lesson that was hard for us to learn, it bears repeating. Our conversations used to go somewhat like this:

> *Potential Vegetarian (PV)*: Oh, so you're a vegetarian. I know someone else who is vegetarian. You know, I really think it's terrible how they treat the animals, but I

could never do it. Animal products are in everything, aren't they?

Vegetarian: They are in a lot of things. But you figure out what you can and can't eat and then it becomes easier.

PV: It's just too much for me.

Vegetarian: I can give you a list of the names of all the different possible animal ingredients. There's fewer than ten thousand of them, all listed in this big book! And I can give you a list of five hundred companies and whether they test on animals or not. It's not so bad. Hey, where are you going?

Now our answer goes more like this:

Potential Vegetarian (PV): Oh, so you're a vegetarian . . . Animal products are in everything, aren't they?

Vegetarian: To me, vegetarianism is not about personal purity, but a way to stop suffering. You don't have to go vegetarian immediately. Just cutting out consumption of chickens and fish will eliminate your support for most cruelty to farmed animals. Have you heard about how horribly chickens are treated on modern farms?

PV: Not really, no, but it just seems like so much. It just seems too hard to do, you know?

Vegetarian: The great thing is, each decision you make is a little victory. Do you think you could eat fewer chickens?

If the person insists on trying to direct the discussion toward animal products and ingredient lists, we can gently and conversationally point out that even if you choose to adopt a vegan diet, there's not much value in avoiding every possible animal product, just the obvious ones for which an animal was bred and eventually killed. Some vegetarians avoid all they can as a symbolic gesture or because they're disgusted by the idea of consuming any animal products intentionally, but as the meat, dairy, and egg industries fade away, so will the minuscule amounts of animal products or by-products.

The best part of veganism isn't about showing how much you give up. Rather, it's about reducing as much suffering as possible—embracing ethics and living positively with compassion. If you're vegan to make a better world—rather than to prove you are "superior" and "pure"—being vegan has to include optimal advocacy. If your choice to worry about some obscure ingredient gives someone an excuse to write-off vegetarianism, you're actively harming animals.

Sometimes a potential vegetarian will say, "I could just never give up my steaks." Now we reply, "Then give up everything but steak, and choose vegetarian every other time. By not eating chickens, you're preventing a lot of suffering." Reactions like this will often surprise the potential vegetarian and make them realize this issue isn't our strict dogma. People often open up to new ideas when they realize they can take steps, and that they don't have to be vegan or nothing.

In the past, when we'd hear: "I couldn't give up X dairy product" (ice cream, cheese, and milk chocolate especially), we used to try to convince the person about the evils of the dairy industry, so we generally lost them totally. Now we remember the numbers (thirty-six birds, one-thirtieth of a dairy cow) and suggest they move incrementally.

Related to this is the importance of finding common ground and praising people when they want praise (for example, when they say "I don't eat much meat," etc.). In situations like this, some activists ignore this opening and immediately set forth trying to get the individual to stop eating all meat (or go vegan NOW!). We want people to work toward vegetarianism, of course, but we will be better advocates if we start with praise. "That's really great. How's it going for you? What kind of vegetarian meals do you like?" Listen to their answers and use their dedication to encourage them to do more.

Using Food to Advocate for Animals

Sharing vegetarian food with non-vegetarian friends is a great tool! Without a doubt, one of the best ways to advocate vegetarianism effectively and to help people have a change of heart about eating animal products is to turn them on to vegetarian food they wouldn't have otherwise encountered.

When presented with information about the nightmarish conditions and cruel practices of factory farming and the potential healthfulness of a well-planned meatless diet, many people can easily understand the "whys" of becoming a vegetarian. It's the "hows" they most often struggle with. When we seek to provide advice for "how," we need to do so from the perspective of a non-vegetarian, not from our own long time experience.

It would be nice if everyone were to watch *Meet Your Meat* or read a booklet (e.g., Vegan Outreach's *Why Vegan?, Compassionate Choices,* or *Even If You Like Meat,* or PETA's *Vegetarian Starter Kit*) and then refuse to eat any animal product ever again—regardless of habit, peer pressure, taste, or convenience. But all experience shows that, for the most part, it's very difficult to overcome rationally or ethically our evolved desire for concentrated protein and fat. Nor can we argue away the need most people have to conform to others around them. A radical change in people's diet from meat to plant foods simply isn't going to happen tomorrow (unless, perhaps, Bruce's dream of the animal rights billionaire comes to pass).

We don't help the animals suffering on factory farms and dying in industrial slaughterhouses when, in addition to a new way of viewing the world, we advocate a strange or exotic diet. Those of us who've been vegetarian for an extended period of time might be perfectly happy with tempeh and quinoa on a bed of organic arugula, but that isn't a recipe for animal liberation in today's America. What's satisfying and savory to us may be repulsive to many who might otherwise be willing to try a vegetarian option for one meal. A tofu-and-carrot burger with nutritional yeast sauce can be enough to sour a meat-and-potatoes eater on anything vegetarian forever, as well as give them a topic of mocking conversation with other meat eaters for the next six months.

This isn't a lesson we learned easily or quickly. For years, we'd make spicy international dishes for non-vegetarian friends and family in an attempt to show how diverse and flavorful a vegetarian diet can be. But all too often, these meals led only to upset stomachs. Replicating "traditional" meals—pancakes and sausage patties, burgers and fries, seitan and gravy with mashed potatoes and white-flour biscuits—almost always works better at leading non-vegetarian friends and family to look forward to our cooking, even for holiday meals.

There are some vegans and animal rights activists who want to package the compassionate vegan diet as one of fresh salads, tofu, bean sprouts, or even all raw foods. Doing so, we feel strongly, is highly counterproductive (unless, perhaps, you live in Berkeley). It's no more compassionate to eat bean sprouts than to eat French fries, or to eat beans or tofu than to eat a veggie burger. We don't need, and shouldn't expect, the masses to give up burgers and milkshakes for tempeh and wheatgrass smoothies. What's needed for a world where animals are not cruelly tortured and barbarically slaughtered for the taste of their flesh is simple: greater demand for vegetarian products that mimic the standard American diet.

There are many great resources available for advancing our food advocacy (see Appendix One: Resources). We need to promote vegetarian products, and share them with our non-veg friends and family. Even if they aren't committed vegetarians, the more people who try vegetarian products, the better. Several people reducing their animal consumption by half saves as many animals as one person going vegetarian. (This ignores the multiplier effect of turning someone into not just a vegetarian, but a vegetarian advocate—but you get the point.) Every vegetarian meal brings us closer to the tipping point, where peer pressure and convenience are no longer pushing people away from making compassionate choices. The tipping point will come at different times for different locations (e.g., San Francisco vs. Omaha) and different populations (college students vs. retirees). But every day, we're closer to the point where faux meats are indistinguishable from flesh. Once the economies of scale kick in, their price will reflect the greater inherent efficiency of vegetarian meats.

Eat With (and Date) Meat eaters

For years, we both refused to eat with meat eaters—it seemed to us too much to be around others feeding on the flesh of our fellows. But we slowly got over our personal discomfort. Avoiding these situations didn't help animals—indeed, it was hurting. Many meat eaters were reading our non-attendance as deprivation, self-righteousness, or both—the sort of club nobody wants to join. ("You can't even go to parties, can't go out to eat! Who wants to live like that?")

One advantage of taking part in gatherings is that people are likely to ask us about what we're eating, especially if they know we're vegetarian. This is a perfect chance to have discussions and get a bit of information into their heads . . . and even into their hands. We try to get a feel for the level of interest; by being respectful, we can often have a good conversation even at a meal where meat is being served, as long as we're upbeat and converse mostly about kindness and against cruelty.

We've heard it recommended that animal activists not bring up vegetarianism at the dinner table—or even that it's acceptable to say that the diet is our "personal preference" and we "don't want to discuss it." This is shocking to us! It's like saying that beating one's dog or torturing one's cat is a "personal preference." Eating meat supports cruelty, and we owe it to animals to be their voice when someone asks us about our diet. It's never acceptable to say that vegetarianism is your "preference," as though it's some odd personal quirk, rather than a moral commitment against suffering.

Sometimes, if it seems like vegetarianism is the last thing people would be willing to discuss at the dinner table, we say something like, "You know, this issue is really important to me. I believe if you saw how animals are suffering on factory farms and in slaughterhouses, you'd be horrified and you wouldn't want to support it. But I've found that having this discussion with a table full of people is often unpleasant for some, and I don't want to monopolize the entire conversation. I'd love to talk with you about this later. Can I get your email address?" Doing this, we've raised the moral issue, even as we alleviate fears that the entire dinner is going to be ruined by a heavy talk. When that person asked, "Why are you a vegetarian?" everyone who expected a long monologue will have warm feelings toward us for choosing not to dominate the conversation. But we will have raised the ethical issue, which is crucially important.

One last thing to say about eating with meat eaters: if we're going to a function where taking food is appropriate, we always take along some tasty, non-exotic dishes (or one "mainstream" and one "exotic"). We also try not to get too angry when all the food we brought gets gobbled up by the meat eaters! We take that as a little victory and, jokingly, point out that people really do like vegetarian food.

Similarly, unless you are going to be totally revolted by the idea of kissing a meat eater, you could consider dating one. You probably have an array of interests other than vegetarianism or animal rights, and you probably used to eat meat. Every person we convince to stop eating meat will be one more person whose choices are as powerful as our own. And what better way to convince people to adopt a vegetarian diet than by discussing it with them while dating them? A friend of Bruce's points out another advantage: if a meat-eating partner goes veg, that shows openness to new ideas—a good trait! And if your new friend belittles you about veganism or animal protection, you'll know they're not the one for you long-term.

Talking with People of Faith

Most of us want to be as effective as we can possibly be on behalf of the animals. Many of us agonize over the perfect answers for every situation. Nevertheless, many in the animal movement seem to have neglected religious outreach, to the detriment of our effectiveness. Think about it: in the United States, nearly ninety-two percent of people identify themselves as members of some Western faith; eighty-six percent are Christians, three percent are Muslims, and three percent are Jewish. Religion often constitutes a crucial part of many people's lives. Even a basic grasp of a few major points may cause someone to pause and reconsider their diet, which may thereby decrease animal suffering.

It is, of course, easy to tire of the common faith-based arguments against animal rights and vegetarianism. We've already touched on one of them, and the others are also familiar: "God put animals here for our use," "What about animal sacrifice in the Bible?" "But Jesus ate meat," and so on. The arguments for faith-based vegetarianism are overwhelming, though (see below, as well as ChristianVeg.com) and are the reason that Bruce is currently involved in animal rights activism. To avoid addressing people of faith is to miss a wonderful and vital opportunity. Here are a few helpful hints and key points for having discussions with people of faith. In all cases, the arguments are similar to the secular arguments with which we're all familiar, but they are presented in a religious context.

- *Don't argue over side issues.* People of faith may want to convert you to their way of thinking, or may be more comfortable/interested in discussing abortion, the death penalty, or the nature of evil. All of these can be interesting discussions, but you can and should keep the discussion focused on the animals killed to be eaten.

- *Find common ground.* Engage people by using statements and concepts with which they already agree (e.g., "animal abuse is wrong," "animals are part of God's creation," and so on). Try not to rewrite the person's scriptures for them—don't argue that the animal sacrifice passages were inserted by meat-eating scribes or that the Gospels proving Jesus' vegetarianism are in a vault under the Vatican. It's not necessary, and you'll be written off.

- *Avoid Bible thumping.* There *is* such a thing as too much information. As with statistics, you can find Biblical justification for just about anything. Biblical support for slavery, murder, and polygamy are actually much stronger than for meat eating. No matter how well you know the texts, people can argue from other perspectives. Even if you have no knowledge of specific religious texts, you can engage people of faith simply by knowing that no religious text makes vegetarianism a sin. General arguments that don't resort to Biblical citation are often more effective and less convoluted, as long as the animal advocate remembers that everyone wants to be viewed as a "good person"—compassionate and thoughtful.

Three Tried and True Suggestions

These specific arguments seem to resonate with people of faith (and you only have to change these suggestions a little for them to work with others as well). This is because they begin with something most people already believe. Virtually none of the rebuttals for these arguments will even begin to adequately answer them. Try to keep coming back to these points, saying, "Well, that's an interesting idea, but I still don't see how it justifies supporting the cruelties of factory farms. . . ."

- God created animals with needs, wants, desires, and species-specific behaviors. God designed pigs to root around in the soil for food and play with one another, behaviors you can see exhibited at sanctuaries for farmed animals. God designed chickens to make nests, lay eggs, and raise their children when they hatch. No less a Christian authority than Jesus compared his love for Jerusalem to a hen's love for her brood. God designed all animals with a desire for sunlight, fresh air, fresh water, and so on. Animals were created to grow at a certain rate that doesn't tax their appendages and organs. Yet, all of these things are denied to God's creatures, who are merely viewed as flesh by modern agribusiness. Animal scientists play "God" by manipulating animals to grow so quickly that their hearts, lungs, and limbs can't keep up. Everything natural is denied them. God's will is negated by the industries that have decided that they know better than God how God's creatures should live.
- Everyone agrees that dogs and cats should be legally protected from the worst abuses. Animal cruelty is regarded as not just unethical, but un-Christian. To their great credit, people of faith fill the mailboxes of the judges in cases of cruelty to animals—when the animals are dogs or cats. But those of God's creation who are raised for food have no protection at all. The

disconnect must be pointed out: if castrating a dog without painkillers is not OK, then it's not suddenly OK because we like the taste of pig or cow flesh. If drugging a cat so that she grows so fast she can't walk isn't OK, if chopping off the toes of a dog isn't OK, if slitting a cat's throat while she is still conscious isn't OK—if any of this isn't OK when done to dogs or cats, it's equally repugnant before God to do these things to any animal. These cruelties are *not* acceptable; God even cares for sparrows.

- People of faith seek to lead moral lives. So it's valid to bring up the issue of paying others to do things they couldn't even watch. All the Biblical justifications for animal slaughter or eating meat fall away when the challenge is issued, "Would you want to work on a factory farm, searing the beaks off of chickens or castrating pigs and cows without painkillers?" Most of us could watch grains being tilled or even spend an afternoon shucking corn or picking beans, fruits, or vegetables. How many of us would want to spend an afternoon slitting animals' throats, or even watching animals having their throats slit? Many people have found the phrase "On Earth, as it is in Heaven" to be useful. How can anyone believe there are factory farms and slaughterhouses in Heaven?

Direct Outreach to People of Faith

If you have a faith background, contact your local clergy about animal issues. Give them literature. Write letters to faith-based periodicals. Consider joining a group such as the Christian Vegetarian Association (ChristianVeg.com, where you can order the booklet, *Are We Good Stewards of God's Creation?*) and the Jewish Vegetarians of North America (JewishVeg.com).

Even if you don't have a faith background, you might want

to let CVA know that you're available to leaflet when Christian conferences come to town; the experience is excellent, and CVA will give you all the tips you need to be successful. Another simple action you can take is distributing literature at places of worship. Pamphlets left in the literature section of churches and synagogues are often picked up and read! Bruce stands outside Christian conventions for entire weekends, showing *Meet Your Meat*, passing out literature, and talking with attendees about vegetarianism. By focusing on cruelty to animals you can reach attendees at a deep and effective level.

Frequently Asked Questions

What follows are some general classes of questions, and some ideas on how to answer them. If you stick with the Socratic method, and remember that no question or concern really answers the basic fact of unnecessary suffering, you should be fine having a discussion with anyone. But here are a few tips for the most common questions and concerns:

Why are you wasting your time worrying about this? Don't you have something better to do? People are starving in Africa!

The first category of question basically expresses the other person's sense that human problems are more important, and thus obviate the need for any concern for animals. They might ask explicitly, "Can't you find something better to do with your energy?" or "Why don't you work on fighting global poverty (or child abuse or abortion)?"

Almost always, the people who ask this sort of question are not fighting global poverty or child abuse in any meaningful way in their lives, so you could easily win the argument by pointing out that the person is a hypocrite. However tempting that might be, your goal isn't to win an argument, but to try to find an effective way of helping this person open his heart and mind. Instead, tell him it's a good question. Point out that you care about humans, too. And bring him around to an understanding that you are simply asking him to live up to his own ethical standards, which will include opposing cruelty to animals.

You may choose to say, "I see what you're saying, and I do support groups such as Amnesty International and Oxfam America that fight for human rights as well. But don't you agree that cruelty to animals should also be opposed?" Once they agree, you might continue by pointing out, "One of the great aspects of helping prevent cruelty to farmed animals is that it takes no extra time. We can continue our activism against AIDS or child abuse while simply choosing a veggie burger instead of chicken flesh at lunch. Here, won't you please read this brochure? I think that it will help explain why this issue is so important to me."

As Peter Singer puts it in *Animal Liberation* (2002):

> Among the factors that make it difficult to arouse public concern about animals, perhaps the hardest to overcome is the assumption that "human beings come first" and that any problem about animals cannot be comparable, as a serious moral or political issue, to the problems about humans. . . . But pain is pain, and the importance of preventing unnecessary pain and suffering does not diminish because the being that suffers is not a member of our species. What would we think of someone who said that "whites come first" and that therefore poverty in Africa does not pose as serious a problem as poverty in Europe? . . . Most reasonable people want to prevent war, racial inequality, poverty, and unemployment; the problem is that we have been trying to prevent these things for years, and now we have to admit that, for the most part, we don't really know how to do it. By comparison, the reduction of the suffering of nonhuman animals at the hands of humans will be relatively easy, once human beings set themselves to do it.

. . . [T]here is nothing to stop those who devote
their time and energy to human problems from joining the
boycott of the products of agribusiness cruelty. It takes
no more time to be a vegetarian than to eat animal flesh.

. . . [W]hen nonvegetarians say that "human problems
come first," I cannot help wondering what exactly it is
that they are doing for human beings that compels them
to continue to support the wasteful, ruthless exploitation
of farm animals.

But I really just don't care about chickens. I don't care if they are
boiled alive. They're only chickens. Why should I care?

In situations like this, try to think about motivation, and believe
that the other person is reachable. It then becomes easier
to construct a reply. Maybe you used to be that way and can
understand the sentiment. People like to feel heard; ask for more
information rather than just launching into a monologue.

You may choose to say, "Well, I know what you mean. I didn't
used to care about chickens either. Do you care about cruelty
to dogs and cats?" After they reply (they'll usually say yes),
you'll be able to explain how farmed animals are the same as
cats and dogs in their ability to feel pain and to suffer and that
they're individuals who don't want to be intensively confined and
violently killed. You may also want to ask them, "Why do you eat
meat?"

Animals eat one another in nature, so why shouldn't we eat them?
Aren't humans at the top of the food chain?
Aren't humans omnivores?

Again, you may choose to say, "I hear what you're saying, and
I used to feel that way, too. But then I realized that in all other
aspects of our lives, we don't rely on the idea that might makes
right to determine our moral values. Like you, I don't support
murder, even though certain animals do fight territorial battles

to the death. And no ethical person endorses rape, even though some animals rape as a method of procreation. As humans, we have the ability to be ethical and kind, rather than cruel and 'natural.' Really, there's nothing natural about factory farming; these places are about as unnatural as you can get: mass cruelty, mass abuse, mass torture. For example: chickens are bred to grow so quickly that their legs often break. Does that make sense to you?"

Then you can move on, perhaps, to say something like, "Wouldn't you agree that we should have laws to protect dogs and cats from being abused?" If you get their assent on that point, you can point out that farmed animals have no legal protection, that what happens to them would be illegal if they were dogs or cats. Here you grant that the question makes sense, find some common ground in combating the argument with points that will resonate with the other person, and then steer the discussion back to cruelty.

But God put animals here for humans to use as we see fit, right?

From conversations we've had, we can tell you that many people don't say this to be callous. They say it because they honestly believe it justifies their meat eating. You may choose to say, "Yes, I hear that a lot, and religion is very important to a lot of people. Would you agree that God opposes cruelty to animals, and that God approves of those who are kind to all creatures?" They'll agree, and then you can continue with something like this: "Actually, some of my closest friends are Jewish/Christian/Muslim—whatever they are (or "I am Jewish/Christian/Muslim . . ."), and they/we are vegetarians because they're/we're horrified by how badly God's animals are treated. From their/our perspective, God designed chickens to build nests and raise their families; God designed pigs to root in the soil; God designed all animals to breathe fresh air, to play with one another, and so on. God made his creation to experience life and serve as witness to His goodness and compassion. But today, animals are denied everything that God designed them to be and to do. Instead, they're treated like machines, and horribly abused. They are God's creatures, but we're treating them like they're rocks. *We* are playing God, really. And of course, the Bible teaches compassion for animals. The horrible cruelty really does deserve condemnation. Don't you agree that cruelty to God's creatures is wrong?"

Don't argue about whether or not God exists or whether the person's religion is valid. Begin by acknowledging that it's a good question. Get the individual to agree with you that cruelty to God's creatures is ungodly. Don't try to convince them that they should have a new interpretation of the Bible, Torah, or Koran. Meet them on their terms. Raise issues that they will understand and that will resonate with them. And as always, bring it back to cruelty.

But we've been eating animals for thousands of years, right?

You may choose to say, "Yeah, we've been eating meat for a long time, but I'm not sure that's a good excuse for continuing to do so. Up until a hundred years ago, you could legally beat a dog to death, but now that's illegal. Would you agree that making cruelty to dogs and cats illegal was a good idea?" If they agree with that, perhaps you can move on in the discussion to say something like this: "We held slaves for most of our existence as a species; we treated women and children as property, and so on. Don't you agree that wasn't right?"

Always remember: validate the question, make a solid moral argument, and steer them back to a discussion of cruelty. Ask them a question to keep the discussion going, perhaps, "Do you think it's wrong to be a vegetarian?"

CHAPTER FOUR

Our Favorite Ideas for Rocking the World

Wear Your Cause on Your Sleeve

Putting a bumper sticker on your car, bike, and/or computer; wearing buttons and cause-related T-shirts; and having cause-related information and brochures with you are simple yet effective ways of raising awareness about an issue. Jack Norris found that simply wearing pro-animal clothing worked: "When I started at a new university, I wore my Vegan Outreach shirts at least every third day. For months, only a few people said anything to me. Some of them joked with me about eating meat. I didn't act offended, and instead I tried to continue the conversation. Slowly, over time, more and more people asked questions. I tried not to be pushy but offered them a *Why Vegan?* pamphlet when the circumstances were right."

Bruce travels quite a bit and so is often working in coffee shops; he has a "Free Vegetarian Starter Kit" bumper sticker on his laptop. Hundreds of people have asked about the kit, and thousands of people see the bumper sticker—each person who sees the sticker thinks about vegetarianism (a little bit) and people who ask for the kit are certain to read it, with some of them sparing an additional three dozen land animals every year.

Offering people a booklet such as *Even If You Like Meat* or *Compassionate Choices* allows them to learn the realities of modern animal agriculture, without feeling the need to justify themselves to us in an "argument." They can also see that the references aren't from a "biased" vegetarian but rather the industry itself. It's critically important that we have literature with us wherever we go. If you carry a backpack or purse, be sure to have some pamphlets with you at all times. For times when you won't be able to carry pamphlets, PETA has business-card-sized leaflets that you can keep in your wallet. They offer PETA's vegetarian hotline and people can then call for free information. Email Bruce if you'd like a free bumper sticker and/or some free vegetarian starter kit cards, or visit veganoutreach.org. Again, never leaving without literature may seem annoying, but your commitment to this will help the animals.

Commit to Online Activism

We're convinced that the World Wide Web is going to help us reduce cruelty to animals much more quickly than we could without it. One simple idea is to advertise a free vegetarian starter kit in your email signature or link to a powerful video (e.g., meat. org). If you have a personal page (MySpace, Facebook, YouTube, etc.), you can post a link to *Meet Your Meat* with some appealing intro. If you spend time on a blog or other site that allows you to post links and content, you could turn your hobby into activism, reaching thousands of people with powerful video and links to life-changing content, without even leaving your chair. Craigslist, Match.com (if you have a profile, add animal rights links), and other sites (what's popular as we write will be different than what's popular when you're reading this)—the possibilities for reaching people are limited only by your time and creativity.

Get Media Attention

A more difficult but crucial proposition, especially if you're in a large city, is trying to get media attention. If you have a letter to the editor published in the *New York Times*, you'll reach a million people, including thousands of the most influential people in society. The lack of detail and pictures in a letter or article compared to a booklet shouldn't deter you from taking these media opportunities. A wonderful service for keeping up with (and responding to) what's happening in the national media is Dawnwatch.com.

Working with the media may seem daunting at first, but it's actually pretty easy. (PETA can work with you.) You can easily make a big difference in your community in many ways. You can alert your local talk radio stations to your availability to come on to discuss animal issues. If you read your daily paper or watch your local news show, you can contact reporters who show sympathies on animal issues to let them know you exist. You can also send letters to your local paper about animal issues.

Another possibility is to do the occasional demonstration. If circumstances align, these can be easy. (Again, if you're interested, please contact PETA, and you'll be walked through the entire process.) If you live in a city with a daily paper that's not one of the thirty biggest cities in the country, and you're especially active on your issue, it's possible that you could convince the paper to do a feature piece on you and your work to make the world a better place (PETA can pitch this to them).

Leaflet

Probably our favorite form of activism, because it's so easy and so effective, is to offer literature directly to individuals. You can do this anywhere there are people. For example, you might keep track of concerts that come to your town and distribute literature where people will line up on public streets or sidewalks at these venues. Or you can leaflet as the crowd leaves. You can sometimes find spots where you can pass out hundreds of leaflets in an hour.

If possible, it's a great idea to add a table with a TV and VCR to your leafleting—though it does take more effort. We recommend that you show *Meet Your Meat*, which PETA will send to you for free on a looped video or DVD. Activists all over the country do this regularly, from LA to Miami, and they report great success in giving out leaflets and having viewers go vegetarian. If you want to do this and you have any trouble with authorities, call PETA— you have an absolute legal right to set up a TV and DVD player or VCR on public property, though sometimes you have to exert a bit of energy to assert that right.

Don't stay behind the table and wait for people to approach you—you should be in front of the table, passing out leaflets and directing people to your table. Being in front with a stack of literature—perhaps saying, "Check out the free video at our table" as you pass out pamphlets—will make your effort more valuable.

Beyond that, the initial impression is crucial in establishing a dialogue. Displays should be designed to garner attention rather than providing detailed information. They should clearly, simply, and immediately convey the area of concern: with a few large words and/or pictures rather than lots of pictures or text. Televisions are always magnets for attracting people, to whom you can then offer literature. Consider your audience and location when you choose which pictures to display or videos to show. Graphic images of animals being tortured understandably upset children, while teenagers and younger adults are more likely to be moved by these photographs.

But if you don't have the time or ability to organize a table, just get out and leaflet! You don't need to get anyone to join you (although feel free to contact us if you'd like to get started with a veteran leafleter). Leafleting can seem intimidating at first. Early on, Matt was nervous before he'd go leafleting. However, he knew that any discomfort he felt was nothing compared to what the animals go through. Once he leafleted a few times and got a sense of how easy yet powerful it is, the nervousness went away. Other leafleters report the same (see veganoutreach.org/enewsletter/profiles.html). One leafleter said that for months, she didn't believe leafleting could be so easy yet effective, and so continued to try to organize specific campaigns. Once a friend convinced her to leaflet, though, she realized everything she'd heard was true!

When someone asks us what they can do to convince a loved one to adopt a vegetarian diet, we suggest that they not dwell on some specific person they know who is highly resistant to change. Instead, go out and leaflet! You'll do more good in an hour than spending years on many of the people you know. In only one or two hours, you can hand out a hundred or more pamphlets at a college (depending on the size of the school), and 300 to 500 per hour in a busy area of a major city or at a concert.

Based on follow-up surveys and feedback received during repeat leafletings at the same location, our experience is that leafleters convince somewhere between three and ten people to change their diet for every three hundred booklets handed out. So in just one leafleting session, you can accomplish three to ten times more good for the animals than you will with every single choice you make for the rest of your life!

A variation on leafleting is leaving literature at sympathetic venues. Do you frequent a particular coffee shop, library, student union, music or bookstore, or other venue that might allow you to keep a stack of *Why Vegan?* or *Vegetarian Starter Kit* brochures by the cash register or in some other place? The great thing about this is that only people who are interested will take them.

Vegan Outreach has an "Adopt a College" program (discussed below) that involves volunteering to leaflet at your local college or university. If you find it especially rewarding, you can join the growing number of activists who are leafleting schools in your area. College students tend to be particularly receptive to new ideas and many of them are interested in becoming more active. Passing out literature on vegetarianism on a college campus can be an exceptionally effective form of activism.

If you're in college yourself, you have even more opportunities to speak for the animals! In addition to leafleting, you can join your campus newspaper's staff and write articles about animal concerns and other important issues, as well as opinion pieces about why one should adopt a vegetarian diet. You can also petition your cafeteria for excellent vegetarian dining options (contact PETA for help in this regard). You can commit to tabling and leafleting on a regular basis to raise campus awareness about vegetarian and animal issues. And you can bring animal advocates to speak on campus and show animal rights videos.

Vegan Outreach's "Adopt A College" Program—i.e., The Best Place to Leaflet

As mentioned earlier, animal advocates have incredibly limited time and resources. Maximum change—the greatest reduction in suffering per dollar donated and hour worked—will come from using our time and resources to present the optimal message to our target audience. This leads to two basic questions: who is our audience, and what is the message that will elicit the greatest change?

We believe that the goal of maximum change points to a focus on high school and college students for three main reasons.

1. The relative willingness and ability to change.

Of course, not every student is willing to stop eating meat. But *relative to the population as a whole*, college and high school students tend to be more open-minded—even rebellious against the status quo—and in a position where they aren't as restricted by parents, tradition, habits, etc.

Most people's behavior is determined by the norms of those around them. If we are raised to be racist, or Christian, or vegetarian, that is generally how we end up. If our peer group changes, though, it's often possible for us to consider new ideas. So the best people to reach are those whose peer groups are in flux, and who are thus most likely to reconsider their worldview and their habits. From social science research, as well as from our own experience, this audience is high school and college students. Reach them before high school, and they're generally too dependent on their parents. Reach them after college, and they tend to have already selected their surroundings, friends, and political views.

2. The full impact of change.

Even if it were equally likely that we could convince teenagers or senior citizens, not only would the teenagers be saving more animals over the course of their lives (because they have more meals ahead of them), but they'd have more opportunities to influence others.

3. The ability to reach large numbers.

High school and especially college students are typically easier to reach in large numbers. For a relatively small investment of time, an activist can hand a copy of *Compassionate Choices* or *Even If You Like Meat* to hundreds of students who otherwise might never have viewed a full and compelling case for compassion. And if you're in college, you should be able to get permission to set up a table and show a video, like *Meet Your Meat*.

Millions of students graduate every year without having been offered a brochure detailing the hidden reality of factory farms and information about the compassionate alternative. We cannot afford to miss this great opportunity to reach college students. Non-students in the animal protection movement must make it our responsibility to ensure that young people are reached while the opportunity exists. If every college in the country were leafleted for thirty minutes tomorrow, there'd be *thousands* of new vegetarians overnight! If you can reach college students who are excited about vegetarianism, please work with them to submit opinion pieces to their school paper and set up a table as people gather for lunch in the cafeteria. If you can get a campus ally, your efforts can be even more effective.

Spreading vegetarianism is a numbers game. The more people you reach with information, the more people will change their diet, and the fewer animals will suffer on factory farms and die in industrial slaughterhouses. By convincing just a few people every semester, you'll be preventing the suffering of thousands of animals over their lifetimes. You'll also be creating a pocket of change that will almost certainly extend out from those people. As the numbers grow, people who have given up eating animals will be much more likely to be sympathetic towards other animal issues; this will increase political pressure in favor of animal liberation.

Even if you have a weekday job, don't discount taking a long lunch or leafleting in the evening (many colleges still have significant pedestrian traffic in the evening). Just thirty minutes on the campus will allow you to reach thirty to fifty new people. Local community colleges count, too; many activists have found community college students to be quite responsive.

Weekdays before three o'clock in the afternoon are normally the busiest times on campuses. Most large universities provide a steady flow of pedestrians throughout the day. Smaller colleges usually have a steady flow between classes, especially around the lunch hour. Some schools, including many community colleges, continue to have a flow of traffic past five o'clock. You might want to drive around any possible school to check out the pedestrian traffic in the evening.

Many organizations and individuals—from clubs to bars to preachers—offer flyers on campuses, and most students are accustomed to being approached. This familiarity can make some resistant to all leaflets, though, regardless of the issue. People often decide whether to take a brochure based on whether the person in front of them took a brochure. If you get a string of individuals who turn you down, it might be wise to stop for about ten seconds and then start back up again. You can stand in one spot or walk around offering booklets to people you come across. Some have leafleted inside academic buildings and student unions when the weather is bad. One extroverted activist actually goes into cafeterias and walks from table to table handing out brochures!

What to Say

Different activists have found the following phrases to be effective:

- *Brochure* (or "pamphlet" or "information") *against* (or "about") *animal cruelty?*

- *Information about vegetarianism?*
- *Help animals?* Or: *Brochure to help animals?* Or: *Want to help animals?*
- *Brochure against factory farming?*
- *Information on non-violent eating?*
- *Info about where your food comes from?*

Feel free to experiment. We don't know why, but we've found that some days one phrase will work very well, and other days another phrase works better.

One long time leafleter, Joe Espinosa, stands in one place and as people pass he loudly exclaims, "Brochure on animal cruelty?!" He gets people's attention and exudes a confidence that implies people should care about animal cruelty. Many people who hear Joe approach him for a brochure, and everyone who passes sees a normal-looking guy speaking out against animal cruelty—that in itself normalizes the issue and is a small victory for animals.

Smiling and being upbeat tend to increase the receptivity rate. Saying "thank you" to individuals for their time, even if they don't take a brochure, encourages many to come back and ask for one. Politeness, friendliness, sincerity, and humility all help encourage people to take a brochure and ask questions. If you dress casually, carry a backpack of pamphlets, and simply walk around approaching students, you will appear to fit in. Even if you're older, people might assume you are a graduate student (or a faculty or staff member) if you dress like one.

Some schools allow leafleting by outsiders, while others do not. According to federal court decisions, public universities are supposed to allow leafleting, but some do not follow such rules; others try to limit leafleting by requiring you register and restricting where you can stand. We are rarely questioned, especially when we just walk around handing out pamphlets, as opposed to standing in one place. Even if someone eventually tells you that you're not permitted to leaflet, it will likely be after you've given out a great number of brochures. Many schools within cities provide a flow of students on public sidewalks where they can be reached. (See veganoutreach.org/colleges/ for more information, including specifics regarding legal issues.)

Kathryn Kovach wrote us after leafleting at Clemson University in South Carolina for the first time: "Thanks for all your help. I needed your pep talk. I handed them out as I walked among the students. Some people ignored me, others laughed as I walked away after I handed it to them, but for the most part it went smoothly. My nervousness did fade as soon as I started handing them out, like you said. And I did see students reading them!"

There will almost always be some students who are glad you're there and who are excited to get the information. For just a small investment of your time, you'll have changed their lives forever!

CHAPTER FIVE

Is Animal Liberation Possible?

> One of the great liabilities of history is that all too many people fail to remain awake through great periods of social change.—Dr. Martin Luther King, Jr.

Even with a sense of history, we know it's possible to become discouraged when we're watching videos, reading about a specific instance of sadistic cruelty to animals, or listening to rationalizations from meat eaters. When reviewing new factory farming techniques in industry trade journals or watching hours of new footage for possible inclusion in the video *Meet Your Meat*, it would be easy to become despondent.

It's also easy to dismiss as naïve the hope for a vegetarian world, given that vegetarians are currently a small (but growing) minority, and that unfathomable cruelty runs rampant and mostly unquestioned on factory farms. Nevertheless, we're absolutely convinced it's possible to achieve our goals—and much more quickly than many of us imagine.

If we look at the long arc of history, we see just how much society has advanced in just the last few centuries. It was over two thousand years ago that the ideals of democracy were first proposed in ancient Greece. But it was only during the eighteenth century that humanity saw the beginnings of a truly democratic system. Not until late in the nineteenth century was slavery officially abolished in the developed world. In all of human history, *only in the last hundred years* was child labor abolished in the developed word, child abuse criminalized, women given the vote, and minorities given more equal rights.

Prejudices we can hardly fathom today were completely

accepted just decades ago. For example, if we read what was written and said about slavery—fewer than one hundred fifty years ago—the defenders were not just ignorant racists, but admired politicians, civic and religious leaders, and learned intellectuals. What is horrifying to us now was once respected. As the Rev. Dr. Andrew Linzey puts it in *Animal Theology*, "[G]o back about two hundred or more years, we will find intelligent, respectable and conscientious Christians supporting, almost without question, the trade in slaves as inseparable from Christian civilization and human progress." In other words, God said we could do it, we've always done it, it's natural. Sound familiar?

However slowly we may feel we're progressing today, we're advancing at lightning speed in comparison to past social justice movements. Just a short century ago, almost no animals received any protection *whatsoever* from abuse. Today the vast majority of people are opposed to cruelty to animals; this really is a significant accomplishment in and of itself! The discussion now must focus on helping people see that eating animals violates their own principles. This effort is only just beginning. If we look back twenty years, most animal advocacy in the U.S. was focused on fur and vivisection—important areas of horrible suffering, but not areas where each individual can have as much of an impact as they can by simply choosing and promoting vegetarianism.

Animal activism in the developed world has never been stronger or more effective than it is today, and we're getting more and more effective each and every day. Two from among many examples: We now have good videos and the ability to share them like never before, and we're reaching out to youth in a way that's unprecedented (as discussed in the previous chapter). Both of these very new developments are cause for optimism.

The Internet has created an advertising and outreach opportunity we couldn't have dreamed of when we started our advocacy. We can put the images of factory farms and solid information about modern farming cruelties directly into people's consciousness like never before. *Meet Your Meat*, perhaps the most powerful tool for convincing people to oppose factory farms, was put together in 2002; in 2007, it was viewed more than one and a half million times online. PETA's youth outreach department was created only in 2002. Vegan Outreach's "Adopt a College" program, the first systematic national effort to reach our most receptive audience, was only officially launched in 2003.

In large part because of our reasoned shift to vegetarian advocacy and our ability and focused efforts to reach new people, factory farms—unknown to most people only a decade or two ago—are now commonly vilified as abominations. Twenty years ago, few people had heard the word "vegan." Finding mock meats and soymilk was nearly impossible. According to market research by Mintel, "Until the mid-1990s, change was slow in coming to the world of vegetarian foods, and many average consumers relegated 'vegetarian products' to a countercultural movement, not a mainstream trend." Today, most people in the U.S. know what "vegan" means, and you can find soymilk, veggie burgers, and various other vegetarian convenience foods in most grocery stores. According to Mintel, "In 2003, the vegetarian foods market in the U.S. topped $1.6 billion in sales. This represents a constant-price growth rate of 111.3% since 1998." Mintel estimates that the market was up to $2.8 billion in 2006.

In addition to the growth in the market for vegetarian foods and regular mentions of vegetarianism and veganism in popular culture, ARAMARK (a company that provides foods on college campuses) reported in 2005 that fully one-quarter of college students valued having vegan meals! ARAMARK has since added 220 vegan and vegetarian recipes to their lineup. *Forbes* magazine reports: "Market research shows that the number of consumers who lean toward some sort of vegetarianism is increasing across all age groups. The Vegetarian Resource Group estimates that 2.8% of adult Americans consider themselves vegetarian, up from 2.3% in a 2000 survey. Another six to ten percent of the population said it was 'almost vegetarian' and another twenty to twenty five percent are 'vegetarian inclined,' or intentionally reducing meat in their diet, according to VRG." According to Food Systems Insider, "Ten percent of twenty-five to thirty-four-year-olds say they never eat meat." (Links and citations for this section at veganoutreach. org/enewsletter/possiblefuture.html.)

As we continue our efforts, more vegetarian products arrive on the market. Having convenient vegetarian options available is vital, as it makes it easier for new people to try *and stick with* a compassionate diet. As more people sample *faux* meats and other vegetarian products, competition will continue to increase the supply and varieties, improving quality and driving down prices. This cycle of increasing numbers of vegetarians and the increasing convenience of vegetarian eating creates a feedback loop that accelerates progress. Essentially, the technology of vegetarian meats and other foods is both driven by and a driver of moral progress.

If we continue to expand our advocacy, the growth of vegetarianism will accelerate to a tipping point, where opposition to factory farms and vegetarianism becomes the "norm" among influential groups. Legislation, as it usually does, will continue to follow these evolving norms, and we'll see more of animal agriculture's worst practices outlawed and abolished—something that's already begun. Corporate practices will also continue to adjust to the demands of an increasingly aware market.

At the same time, powerful economic forces will kick in, because meat is, ultimately, inefficient. Basic biology makes it inherently more efficient to eat plant foods directly, rather than feeding crops to animals and then eating some of the animals' flesh. This doesn't mean that capitalism favors a plate of soybeans—or even tofu—over a rump roast or chicken leg. However, that's not the choice people will face. Food science has advanced so much that the best vegetarian meats are able to satisfy even hard-core carnivores. Deli slices from Tofurky, burgers from Boca, Gimme Lean sausage and ground beef, Morningstar MealStarters, Gardenburger's Riblets and Chik'n—all of these belie the notion that eating vegetarian is a deprivation. The faster the growth in the number of people eating vegetarian, the faster vegetarian meats will improve in taste, become cheaper, and be found in far more places. Our challenge now is to expand the vegetarian market by explaining to more meat eaters the reasons for choosing vegetarian meals, while exposing them to new yet familiar-seeming products. The more rapidly we do this, the sooner cruelty-free eating will be widespread.

Imagine that there are just 50,000 vegetarians right now (there are actually millions) and that each of these 50,000 convince just one person to stop eating animals over the next five years. Imagine that *those* 100,000 convince just one person in the five years after that, and that those 200,000, etc. In fewer than seventy years, we have a vegetarian America, even accounting for population growth. Most of us will do better than influence one person every five years—some of us will be able to open the hearts and minds of hundreds of people every year. The harder we work, the faster we achieve that vegetarian America we're striving toward.

Despite all the current horror and continued suffering, if we take the long view *and are willing to commit to the work that needs to be done*, we should be deeply optimistic. If we take suffering seriously and are committed to optimal advocacy, we can each create real, fundamental change. Animal liberation *can* be the future. As the magazine *The Economist* concluded, "Historically, man has expanded the reach of his ethical calculations, as ignorance and want have receded, first beyond family and tribe, later beyond religion, race, and nation. To bring other species more fully into the range of these decisions may seem unthinkable to moderate opinion now. One day, decades or centuries hence, it may seem no more than 'civilized' behavior requires" (August 19, 1995).

It's remarkable to recall that just three hundred and fifty years or so ago, the pope sentenced Galileo to the torture chamber until he would recant the "heresy" that Earth isn't the center of the physical universe. With our efforts, society will recognize that humans aren't the center of the moral universe. In the not-too-distant future, society will look back on the subjugation of animals with the same sort of horror and disdain that we currently reserve for past atrocities against human beings. This century will be the one in which society stops torturing and slaughtering our fellow earthlings for a fleeting taste of flesh.

Ultimately, *de facto* animal liberation will be achieved, but not with a bang. Change will not come by revolution, but through person-by-person outreach progressing hand-in-hand with advances in technology, leading slowly but inexorably to a new norm that, to most people, hardly seems different. But an unfathomable amount of suffering will be prevented.

It is up to us to make this happen.

Because of the number of individuals suffering and the reason behind this hidden brutality, we believe that animal liberation is the moral imperative of our time. We can be the generation that brings about the next great ethical advance. We should revel in the freedom and opportunity we have, the ability to be a part of something bigger, something *fundamentally good*. This is as meaningful and joyous a life as we can imagine!

We have no excuse for waiting. Taking action against suffering doesn't require anything other than our choice. We can act *today*.

In the end, in our hearts, we know that, regardless of what we think of ourselves, our actions reveal the kind of person we really are. We each determine our life's narrative. We can, like most, choose to allow the narrative to be imposed on us, mindlessly accept the current default, follow the crowd, and take whatever we can. Or we can choose to actively author our lives, determining for ourselves what is really important. We can choose to live with a larger purpose, dedicated to a better world for all. We can choose to be extraordinary!

The choice is fundamental. The choice is vital. *And the choice is ours, today.*

Appendix One

Resources

Rather than offering a lengthy list, we've narrowed our list to just a few items in each category. Visit animaladvocacybook.com for links and more information.

Websites

For animal advocacy, we suggest PETA.org and VeganOutreach. org, for youth activism PETA2.com.

In addition to the Starter Guide section of VeganOutreach. org, VegCooking.com (called by *VegNews* magazine the best vegetarian-cooking site online) is a great tool to show people how easy being a vegetarian can be. Full of recipes and cooking tips, this informative and insightful website serves up everything needed for an exciting and diverse diet and a humane and healthy lifestyle. Whether you're looking for a vegetarian version of a favorite recipe or want to prepare something a little more exotic, VegCooking.com's database of over 1,500 kitchen-tested recipes (with more being added every week) offers something for everyone. There's even a two-week sample menu plan to use as a guide, lots of great vegetarian cooking tips, cookbook recommendations, and an interactive "Ask the Vegan Chef" section with cookbook author Robin Robertson.

People commonly think that to become a vegetarian, you have to change your entire lifestyle. But of course you don't have to change anything except some parts of your diet. VegCooking. com accommodates the lifestyle that carnivores know and love by providing a variety of seasonal recipe features that not only help them to make the best use of fruits and veggies when they're at their freshest, but also lends a hand in the kitchen for holiday and special event cooking. There are s'mores for a camping trip, cool treats for a day at the beach, and stick-to-your-ribs recipes for a Fourth of July or Labor Day barbecue. Vegetarian versions of hundreds of other seasonal favorites are found in VegCooking. com's archives. There are also time-saving features, such as an expansive shopping list of easy-to-find products and familiar brands that are vegetarian, a "New Products" section highlighting the latest and greatest vegetarian food products on the market, and a section that explains how to substitute common animal derived ingredients in recipes and provides alternatives for meat, dairy, and eggs.

Books

We suggest the Web for the best and most up-to-date information on vegetarianism (especially GoVeg.com and VeganOutreach. org), but for books on effective conversations and advocacy, we suggest the books referenced throughout the text, especially Dale Carnegie's *How to Win Friends and Influence People* (1990), Malcolm Gladwell's *The Tipping Point* (2000), and Dr. Robert Cialdini's *Influence* (2006). And for time management, we suggest David Allen's *Getting Things Done* (2003), Steven Covey's *The Seven Habits of Highly Effective People* (1990), and Julie Morgenstern's *Never Check E-mail in the Morning* (2005). For a brilliantly insightful discussion of ethics in general, we recommend Peter Singer's *Writings on an Ethical Life* (2000), which has excerpts from the great books *Animal Liberation* (2002), *Practical Ethics* (1993), and *How Are We To Live?* (1995).

Authors' Note: We've done a lot of debating as we were putting this book together, and our thoughts are always a work in progress. If you find something in here that you disagree with, some nuance that you think could improve our thinking, or if you read a book that you think would help us to improve our arguments regarding how to effectively bring about change over a lifetime, please don't hesitate to let us know.

Bruce can be reached at BruceF@peta.org, and Matt at Matt@ VeganOutreach.org. If you don't have internet access:

Vegan Outreach
PO Box 30865
Tucson, AZ 85751

PETA
501 Front St
Norfolk, VA 23510

Appendix Two

Humane Meat?

More and more people are becoming aware of the terrible mistreatment suffered by virtually all of the ten billion land animals slaughtered in the U.S. each year for their meat, eggs, and milk—and that fish farming and commercial fishing also cause much suffering. In fact, routine farming practices are so abusive that they would warrant felony animal cruelty charges were they done to cats or dogs.

As a result, huge numbers of compassionate people have joined the ranks of the vegetarians. Some, however, have instead looked for meat from animals treated less badly, which they sometimes call "humane meat." Some people argue that promoting "humane meat" would better serve animals than vegetarian advocacy.

But does truly "humane" meat exist? And would consuming only "humane" meat conform to an ethical life?

Before addressing these questions, it's important to clarify a few points. First, our goal is not to promote vegetarianism for the sake of promoting vegetarianism; rather, our goal is to prevent as much suffering as possible, now and in the future. Our combined experience, as well as our study of other efforts, tells us that nothing even comes close to serving that goal as well as vegetarian advocacy.

However, we realize that many people will continue to eat animal flesh, eggs, and milk for many years, no matter how hard we work right now. Hundreds of billions of animals will be raised and slaughtered to be eaten before worldwide vegetarianism is established; therefore, we seek to lead as many people to as much change as possible. It would be great, for example, if everyone were to eat fewer animals, even if they don't go vegetarian right away (or ever). This would lead to many fewer animals suffering on factory farms and being slaughtered in industrial slaughterhouses; several people eating half as much meat will help the same number of animals as one vegetarian, and so on. Someone who gives up eating fish, chickens, eggs, and turkeys is removing their personal support for more than ninety-five percent of their previous support of cruelty toward farmed animals, as discussed previously.

Furthermore, we recognize that not every animal product is the product of equally cruel production, nor does every animal product support the same amount of exploitation and suffering. For example, not only does one serving of chicken flesh have more suffering behind it than one serving of dairy, dairy production in this country varies widely, from small, free-ranging herds in temperate climates (the image most people have of dairy farms), to huge industrial production facilities, where the cows never go outside. Sometimes, the cows can't even walk around. Some dairies are even located in the desert! While some activists equate any form of exploitation as equally wrong, we realize that certain farms and production methods are crueler than others, and we support all efforts that truly reduce suffering in the real world.

This is why we believe that activists should support (even if they don't focus on) changes to the system, like bans on gestation and veal crates, the campaigns against McDonald's, Burger King, and KFC, etc. (We recommend the essays on this topic that are posted at AnimalAdvocacyBook.com.) We also believe that admitting the differences in levels of cruelty shows people that we aren't extremists trying to promote our personal, absolutist dogma, but rather, we're truly focused on the animals and their real suffering.

To be effective advocates for the animals, we must honestly evaluate the world as it currently is, and then do our very best to reduce as much suffering as possible. We seek to reach and influence the people who might be willing to go vegan; reach and influence the people who might be willing to go vegetarian; reach and influence the people who won't (now) go veg, but who might stop buying meat from factory farms—and help support all of these people as they continue to evolve. Outreach efforts to all of these people are necessary if we're to help a large and diverse society evolve to a new ethical norm. This is why, for example, Vegan Outreach produces a range of literature to make everyone and anyone, in any situation, the most effective advocate for animals possible. While we're each able to do this person-to-person outreach in our areas, we support—and certainly don't waste our limited time *opposing*—the efforts of large organizations to bring about the abolition of the worst abuses on factory farms. Each step brings the animals' interests to light, helping some people consider the otherwise hidden reality behind the meat they eat. There's simply no other way to go from a carnivorous society, where farmed animals have virtually no protection, to a vegan society where they have near-total protection.

Animal agriculture certainly recognizes the recent changes in society's awareness. However, agribusiness doesn't seek truth in advertising. Rather, it's happy to play on consumers' concerns. Indeed, many of the "humane" labels, such as "Swine Welfare Assurance" and "UEP Certified," are basically meaningless. Animals raised under these guidelines are treated in the worst imaginable ways—all the standard abuses are permitted. (You can find a guide to the labels at GoVeg.com, in the "organic and humane" section.)

Even so, at first blush, it may seem easier to eat "humane meat" than choose vegetarian, but in fact it's not—products with relatively legitimate, meaningful, enforced labels are far harder to find than vegetarian foods. Indeed, for all intents and purposes, if you can't visit a farm and see how the animals live and die, you really can't be sure of how "humane" they really are treated.

For example, in *The Omnivore's Dilemma* (2007), Michael Pollan endorses eating animals from Pollyface farm, where "animals can be animals," living, according to Pollan, true to their nature. But what is Pollyface really like? Rabbits on the farm are kept in small suspended-wire cages. Chickens are crowded into mobile wire cages, confined without the ability to nest or the space to establish a pecking order. Pigs and cattle are shipped year-round in open trucks to conventional slaughterhouses. Seventy-two hours before their slaughter, birds are crated with seven other birds. After three days without food, they are grabbed by the feet, up-ended in metal cones, and, without any stunning, have their throats slit.

This is the system Pollan proclaims praiseworthy. In the end, Pollyface's view is the same as Tyson's—these individuals are, ultimately, just meat to be sold for a profit. It's logically and emotionally impossible for there to be any *real* respect, any *true*, *fundamental* concern for the interests of these individuals when these living, breathing, feeling animals exist only to be butchered and consumed. If we insist that we must consume actual animal flesh instead of a vegetarian option, it's naïve, at best, to believe any system will *really* take good care of the animals we pay them to slaughter. If you say an individual is just meat, they'll be treated as such.

From an advocacy perspective, those of us in a position to influence major campaigns can debate, for example, whether a slaughter practice is more or less humane than another. But it's important to remember the underlying point—the sheer outrageousness of harming and killing another animal without good reason. We have no nutritional need for meat, eggs, or milk (if you have any doubts, see veganhealth.org/articles/realveganchildren for a page of thriving, lifelong vegans). Eating meat means, quite literally, eating a corpse. It means robbing an individual of her or his life, and then devouring the body. Consuming or promoting "humane" meat is not a different diet, nor an ethical example, nor an effective advocacy technique. It's just a variation on the view of animals as meat. Calling any flesh-food "humane"—a word that references the very best of our nature—bastardizes the very idea.

Albert Einstein and Leo Tolstoy pointed out that the fundamental essence of morality is how we use our power in relation to the weak and innocent. Tolstoy summed it up by saying, "Vegetarianism is the taproot of humanitarianism." Einstein spoke of the human arrogance that considers us superior to other species, calling this rationalization for exploiting our fellow creatures "a kind of optical delusion of consciousness." He knew that "our task must be to free ourselves from this prison by widening our circle of compassion." He is right. To be truly free, we must move beyond the current ethical norms of our society. How can we claim to be a good person while dining on the corpse of a defenseless victim?

In the end, it really is a question of what kind of person we choose to be. Do we oppose cruelty or support slaughter? Do we make our own decisions or do we rationalize what we're used to doing? Only vegetarianism makes an unalloyed statement against cruelty at every meal. It's the only choice that fits in with a consistent, ethical narrative of life. It's incredibly fulfilling to know that you are promoting kindness rather than cruelty, affirming life rather than creating death, and helping to bring about a society that is life-giving rather than life-taking.

Appendix Three

A Theory of Ethics

A version of this essay is at: veganoutreach/advocacy/theoryofethics. html

Most people believe that adult humans are moral beings, capable of acting not from instinct but rather from a reasoned set of rules. These rules are generally called "ethics." For most of history, discussion of "ethical rules" was dominated first by superstition and later by religious doctrine. Because most people weren't able to read, they were reliant on leaders (tribal, religious, etc.) to tell them what to believe. Thus, ethics were largely devoid of serious reasoned examination. Only in the last few centuries have ethics been rigorously pursued outside of religious doctrine. In many countries today, even individuals who hold strong religious convictions are dependent upon arguments from secular ethics to resolve disagreements, and most religious doctrines now accept that their texts should be viewed critically as products, at least in part, of human cultures.

Pope John Paul II stated that religious principles must hold up to rational ethical argument if they are valid. In other words, he granted that no one should take any ethical principles "on faith." Any student of theology knows that religious texts were clearly created by humans in a specific time and place. Whatever we take from these writings will require historical and ethical analysis, rather than an uncritical read. It's a good thing, too, or what would we do with sections like these?

When your brother is reduced to poverty and sells himself to you, you shall not use him to work for you as a slave. . . . Such slaves as you have, male or female, shall come from the nations round about you; from them you may buy slaves. You may also buy the children of those who have settled and lodge with you and such of their family as are born in the land. These may become your property, and you may leave them to your sons after you; you may use them as slaves permanently (Leviticus 25: 39–46).

While the Israelites were in the desert, a man was discovered gathering wood on the Sabbath day. Those who caught him at it brought him to Moses and Aaron and the whole assembly. But they kept him in custody, for there was no clear decision as to what should be done with him. Then the LORD said to Moses, "This man shall be put to death; let the whole community stone him outside the camp." So the whole community led him outside the camp and stoned him to death, as the LORD had commanded Moses (Numbers 15: 32–46).

There are many similar examples from the Bible, used for millennia to justify the Crusades, witch burnings, slavery, child abuse, denial to women of the right to vote, condemnation of homosexuals, and so on. In every case, the Bible was used to justify the wrong side of history, as admirably discussed in Ernie Bringas' exploration of the issue, *Going by the Book: Past and Present Tragedies of Biblical Authority*.

If a Jew or a Christian claims their religious texts are an infallible guide to morality, they would have to say that slavery is ethically acceptable, working on the Sabbath warrants the death penalty, and so on. Alternatively, if one does not consider a literal reading of a religious text as the first, last, and only word on ethics, then one is left to find another, rational basis for ethics. To reduce the problem of interpretation and the prevalence of inherent prejudices, one might do well to seek a universal basis that can transcend the boundaries of faith and culture. Of course, seeking such a universal basis is not at odds with developed religious belief; any Creator who has given humans the ability for rational thought and logical analysis would want us to use these abilities.

Despite the capacity for rationality, human beings have several significant obstacles to overcome when discussing ethics. Foremost, we have significant evolutionary baggage that leads us to value ourselves and family first, our tribe second, and strangers third—if we value them at all (Druyan and Sagan, 1993; Wright, 1995). Some people call this hierarchical value system our "moral intuition," or our "moral instinct"—what "feels" or "seems" right *is* right (i.e., ethical). Some philosophers derive ethics from our instincts and intuitions. Intuitionists may judge ethical arguments against our intuitions and modify these arguments so as to fit better with our intuitions. Still other philosophers start with their intuitions and work backwards to try to create some seemingly rational basis to justify their desired conclusions.

Not everyone has the same instincts about ethics, however, and many instincts are contradictory, and thus cannot all be valid. Indeed, we now widely condemn as unethical the instincts of those who enslaved blacks, burned witches, forced children to work in mines, and so on. We are convinced that a century from now, people will condemn our generation for instincts we may now be harboring uncritically. It is important that ethics, whenever possible, avoid deferring to potentially prejudiced instincts.

Universal Ethics

When discussing ethics, the easiest means of avoiding our prejudices is to take as objective a point of view as we possibly can. Such a point of view is sometimes called "the point of view of the universe," a view that allows us to empathize with all those beings affected by a decision. One of the more common methods of approximating this view is called "The Original Position." Imagine yourself as a disembodied entity, existing outside the world. At some unknown time in the future, you will be "incarnated" on Earth, at which point you will take on the intellectual and emotional characteristics of your new body. In addition, you do not know your future IQ, your race, your nationality, your gender, or even your *species*. (Although most philosophers don't include species variation in this calculation, they haven't given a good reason for species to be excluded.)

Behind this "veil of ignorance," (a concept created by Harvard philosopher John Rawls), you must choose what is to be held good and bad in the world in which you will be incarnated: i.e., what rules of ethics should be followed. Because you are self-interested, you want to protect whatever interests you may have in your various possible incarnations. Put another way, a universal view like that of the Original Position involves an "equal consideration of interests" of all those beings one could become.

How can one think about a situation like this? What can be said about the various beings whose lives we could possibly lead? How can we compare their diverse interests? One universal aspect is that every being said to possess "interests" seems to pursue experiences that they find desirable (pleasure) while avoiding those that are undesirable (pain)—in short, maximizing pleasure while minimizing pain are interests held by each individual with the biological capacity for having interests. Such interests appear to be fundamental to all conscious creatures, likely the result of evolutionary processes that used pleasure and pain as inducements to guide behavior and learning. If organisms (such as bacteria, plants, and most likely some simpler animals like clams and some other invertebrates) are incapable of the subjective experience of pleasure or pain, then the rules by which one interacts with them are irrelevant to *them*. You could be incarnated as an oak tree, but the universal system of ethics set forth would be inconsequential to *you*.

For sentient, conscious beings capable of subjective experiences, these interests vary as widely as the organisms do:

from basic avoidance of nerve tissue damage, to the conscious, intellectual desire for "justice." What seems likely to be universal, however, is that vertebrate animals are aware of pain and pleasure (Bateson, 1992).

Pleasure and pain thus provide a universal basis for ethics in which the interests of diverse beings can be compared. Knowing nothing more than this, one can set forth a basic ethical rule for the world into which they will be incarnated: a conscious being's interests in a pleasurable, minimally painful life will be respected as much as the comparable interests of other beings. In short: equal consideration of interests.

Differences of Interests

Equal consideration of interests does not imply equality of treatment. Individuals have different interests and thus require different treatment to protect these interests. As Richard Ryder points out (using the language of rights), "Humans suffer if denied the right to vote, so this is important for humans but it is not so for other species. Access to eucalyptus leaves is, however, important for koalas, and so the right of access to eucalyptus leaves is an important right for them."

Furthermore, not all interests are of the same intensity. As Bernard Rollin writes: "I would not adopt as a universal principle always favoring the 'higher' animals for example, if the choice came down to a quick death for the higher animal versus a slow, lingering death for a lower animal, one should presumably choose the death of the higher animal. This makes us realize that we need to consider not only *number* of interests, but also quality and intensity of their satisfaction and frustration."

Similarly, our interest in finding pleasure and avoiding pain may not be equal. It is possible for an individual to have a variety of pleasurable experiences, but the range of pleasures in life does not seem to match the range of pains. As Richard Ryder writes, "At its extreme, pain is more powerful than pleasure can ever be. Pain overrules pleasure within the individual far more effectively than pleasure can dominate pain." Some will balk at this, but think of the most pleasurable experience you've ever had. Most people can easily come up with a list of experiences that they wouldn't be willing to endure in order to experience that pleasure (e.g., various forms of torture, watching a loved one die a terrible death, etc.). In other words, there are limitless unpleasant experiences that you would not suffer in order to have even the most intense possible pleasure. All this isn't to say that pleasure doesn't count at all, but that, in general, equal consideration of interests focuses on the reduction and elimination of suffering.

Applications

Once we've arrived at the view of equal consideration of interests and see that pleasure and pain form a common currency with which to compare these interests, we must ask what this means for our ethics. First, it will take us to many of the conventional ethical positions that most of us already accept: suffering is bad; hunger and disease should be alleviated; people should be given personal freedoms with which to prosper; individuals shouldn't be discriminated against on the basis of their race, gender, nationality, sexual preference, or other group membership; laws should protect the interests of the weaker against the stronger.

However, our view of equal consideration also leads us to some conclusions that run counter to current conventional ethics, particularly with regard to non-human animals. If suffering matters, then much of our current treatment of animals is unjustifiable. For instance, we may gain some pleasure from eating a fish or chicken sandwich. However, equal consideration of interests makes us put ourselves in the place of the animal as well as in the place of the sandwich-eater. Does the pleasure of eating a fish or chicken sandwich—*instead* of a veggie burger or other vegetarian option—outweigh the pain we would endure to be raised and killed for that sandwich? If we were incarnated as a chicken or farmed fish, we would conclude that the interest in not being abused and slaughtered is stronger than the pleasure gained by eating a meat sandwich instead of a veggie burger.

Objections

The universality of this "equal consideration of interests" theory of ethics is straightforward. The sole logical, rational, and reasonable manner for building a truly universal ethic is by including everyone with interests, including other animals. What is important is determined only by the nature of, and consequences for, those affected by decisions. Yet instincts and prejudices are far older than formalized ethics, and run as deep as our evolution. Thus, it may be worthwhile to examine some of the objections against universalized ethics. (For a discussion of more objections to these ideas, see "Beyond Might Makes Right" at veganoutreach.org/advocacy/beyond.html)

Excluding Animals: Moral Contracts

One objection is that the pool of possible incarnates includes those unable to act from the chosen code of ethics—such as children, the mentally handicapped, and most non-human animals. In other words, "in order to have rights, you must also have duties." This is a basic "moral contract" theory. Some "contractualists" argue that anyone who cannot also have duties does not deserve direct moral consideration. Is this reasonable?

Given that infants, children, etc., can be affected by the decisions of moral agents, there's no consistent reason for excluding them from the pool of possible incarnates, and thus from consideration of their interests. It would be in the interests of those in the original position to include these states within their code of ethics, since all moral agents begin life unable to participate in any moral contract (as infants and children) and can become that way (after a stroke or senility). As Rollin concludes, "In a nutshell, there is no argument showing that only moral agents can be moral recipients. On the contractualist view, it is also hard to see why animals differ in a morally relevant way from all sorts of humans who can't rationally enter into contracts—infants, children (especially terminally ill children, who will not live long enough to actualize rationality), the retarded, the comatose, the senescent, the brain-damaged, the addicted, the compulsive, the sociopath, all of whom are also incapable of entering into or respecting contracts."

Moreover, it's not even clear that the distinction between moral agents and others would exclude all nonhuman animals. While a handful of adult humans may claim a monopoly on ethical *theory*, humans do not have a monopoly on ethical *practices*. As Drs. Carl Sagan and Ann Druyan relate in *Shadows of Forgotten Ancestors* (1993):

> In a laboratory setting, macaques were fed if they were willing to pull a chain and electrically shock an unrelated macaque whose agony was in plain view through a one-way mirror. Otherwise, they starved. After learning the ropes, the monkeys frequently refused to pull the chain: in one experiment only 13% would do so; 87% preferred to go hungry. One macaque went without food for nearly two weeks rather than hurt its fellow. Macaques who had themselves been shocked in previous experiments were even less willing to pull the chain. The relative social status or gender of the macaques had little bearing on their reluctance to hurt others.

> If asked to choose between the human experimenters offering the macaques this Faustian bargain and the macaques themselves suffering from real hunger rather than causing pain to others, our own moral sympathies do not lie with the scientists. But their experiments permit us to glimpse in non-humans a saintly willingness to make sacrifices in order to save others—even those who are not close kin. By conventional human standards, these macaques—who have never gone to

Sunday school, never heard of the Ten Commandments, never squirmed through a single junior high school civics lesson—seem exemplary in their moral grounding and their courageous resistance to evil. Among these macaques, at least in this case, heroism is the norm. If the circumstances were reversed, and captive humans were offered the same deal by macaque scientists, would we do as well? (Especially when there is an authority figure urging us to administer the electric shocks, we humans are disturbingly willing to cause pain and for a reward much more paltry than food is for a starving macaque [cf. Stanley Milgram, *Obedience to Authority: An Experimental Overview* (2004)].) In human history there are a precious few whose memory we revere because they knowingly sacrificed themselves for others. For each of them, there are multitudes who did nothing.

Discussing the macaque monkeys who chose to starve rather than inflict pain on another, Drs. Sagan and Druyan conclude, "Might we have a more optimistic view of the human future if we were sure our ethics were up to their standards?"

Excluding Animals: Rationality

People are generally willing to include all humans (including babies, the insane, etc.) in their circle of ethics, but balk at a truly universal ethic (i.e., the inclusion of non-human animals in the pool of possible incarnates). This is because of the obvious and difficult implications, notably that one should not eat or generally cause animals to suffer. Many find the consequences of this inclusion unacceptable, and it's to avoid these consequences that the vast majority of philosophers and ethicists have either simply ignored animals or created arguments trying to show that only humans require ethical consideration.

Is it possible to build a rational and morally relevant argument for the exclusion of animals instead of simply including everyone in the Original Position? Despite the efforts of many, it is unclear how one might do this. For instance, in his book *A Theory of Justice* (1999), John Rawls argues that only moral agents are to be included. Rawls attempts to count children among moral agents because they are *potential* moral agents. However, as Singer writes, this is "an *ad hoc* device confessedly designed to square his theory with our ordinary moral intuitions, rather than something for which independent arguments can be produced. Moreover, although Rawls admits that those with irreparable mental defects 'may present a difficulty,' he offers no suggestions towards the solution of this difficulty" (*Practical Ethics*).

What is it about being rational that makes it ethically relevant for inclusion in the set of potential incarnates? Of course, rationality is an assumption for an analysis of the Original Position, for there'd be no discussion otherwise (if irrational decisions were allowed, anything would be fair game, and there'd be no basis for a set of rules governing interactions). Yet making rationality a requirement for being a *potential* incarnate has no basis. If rationality were a prerequisite, many "marginal" human beings (such as the brain-damaged and senile) would be excluded from moral consideration, although there's nothing to indicate that, as biological beings, rationality is inherent even in theory.

Excluding Animals: Intelligence

Intelligence is often offered as the function that sets humans apart from other animals. Rollin counters this common contention: But why does intelligence score highest? Ultimately, perhaps, because intelligence allows us to control, vanquish, dominate, and destroy all other creatures. If this is the case, it is power that puts us on top of the pyramid. But if power provides grounds for including or excluding creatures from the scope of moral concern, we have essentially accepted the legitimacy of the thesis that "might makes right" and have, in a real sense, done away with all morality altogether. If we do accept this thesis, we cannot avoid

APPENDIX THREE   111

extending it to people as well, and it thus becomes perfectly moral for Nazis to exterminate the Jews, muggers to prey on old people, the majority to oppress the minority, and the government to do as it sees fit to any of us. Furthermore, as has often been pointed out, it follows from this claim that if an extraterrestrial alien civilization were intellectually, technologically, and militarily superior to us, it would be perfectly justified in enslaving or eating or exterminating human beings.

Or to quote the founder of utilitarian thought, Jeremy Bentham, in discussing what it is that qualifies one to be worthy of moral concern: "Is it the faculty of reason, or, perhaps, the faculty of discourse? But a full-grown horse or dog is beyond comparison a more rational, as well as a more conversable animal, than an infant of a day, or a week, or even a month, old. But suppose the case were otherwise, what would it avail? The question is not, 'Can they reason?' nor, 'Can they talk?' but, 'Can they suffer?'"

Indeed Bentham's point is well taken, although, more and more, scientists are learning that animals are interesting individuals with rich intellectual lives. We now know that pigs function intellectually at a very high level (by some measures, beyond that of your average human three-year-old), and chickens have certain capacities beyond that of a child of one. Pigs can learn from one another, play video games, and more; chickens have the capacity for foresight, delayed gratification, and the ability to figure out the existence of an object they can't see.

But in practice, humans don't make ethical decisions based on a hierarchy of intelligence. For the same reason that most people bring their family dog or cat into their realm of moral concern—they understand the animal as an individual—there's no reason to exclude other animals either. Or, as Charles Darwin put it, "There is no fundamental difference between man and the higher animals in their mental faculties. . . . The lower animals, like man, manifestly feel pleasure and pain, happiness, and misery."

Excluding Animals: Language

R. G. Frey (1980) argues that only with language can a creature have interests: "If what is believed is that a certain declarative sentence is true, then no creature which lacks language can have beliefs; and without beliefs, a creature cannot have desires. And this is the case with animals, or so I suggest; and if I am right, not even in the sense, then, of wants or desires do animals have interests."

Let's assume for a moment that animals don't have language and see if the argument would hold up to rigorous scrutiny: while quick to use this rationalization against animals, Frey ignores the implication for infants and brain damaged humans. (Furthermore, without beliefs and interests, why and how would infants acquire language?) Still others, such as Michael Leahy, will go so far as to allow for the exclusion of "marginal humans," so as to be able to reject consideration for other animals! Indeed, given that fully matured Broca and Wernke areas of the brain are required for language, if Frey were to have a minor stroke in one of these areas, he would no longer be subject to ethical consideration, and, based on his argument, could be subsequently eaten or used for experiments, regardless of the suffering he experienced. It seems safe to say that Frey has never experienced severe pain, or else he'd know that language isn't required to have interests, as the deepest sufferings often overwhelm the ability to think in language.

This isn't to imply that the brain is not required to have interests. Damage to the brain can lead to the loss of interests—thus the term "brain dead." Even relatively small damage, such as the destruction of the hypothalamus in the case of Karen Ann Quinlan, who lived for many years in a persistent vegetative state, can end one's interests.

What does speaking represent? Does language create an entirely new (inner) world—one alien to or different from those without language (infants, animals)? Did Koko, the ape who learned sign language, become a wholly new creature? At what point in learning a language does an infant have an interest in not being tortured? A bigger question is this: before Koko learned a form of human language, did she perhaps have another form of language, one we couldn't understand? It begs credulity to think otherwise, as we continue to learn about communication in other mammals (including rodents), in birds, and in fish. Dr. Temple Grandin, in her book *Animals in Translation: Using the Mysteries of Autism to Decode Animal Behavior*, points out that science has always started from the premise that other animals *can't* do certain things, and then learned that, in fact, they can. She notes that every time we assume they can't (use tools, plan ahead, learn from one another or from humans, and so on), we discover that we were wrong. If experiments have failed to show as much, it turns out that it was the experimenters who were failing, not the animals.

Excluding Animals: Our Interests Are More Important

One might allow that many non-human animals have interests, but find no inherent implications from this admission. Indeed, few call for outright and total dismissal of animals' concerns (e.g., advocating the repeal of current welfare and anti-cruelty laws). Rather, the current Western consensus is that humans' interests are simply *more important*. But to whom are humans' interests more important? To *humans*, of course. But do we really contend that our potential interest in eating a chicken nugget instead of a vegetarian option is greater than the chickens' desire not to be drugged, cooped in their own waste, transported through all weather extremes, and cruelly slaughtered?

Defenders of animal experimentation often use emotional hypothetical choices of "more important" to defend animal exploitation. For example, concerning her daughter Claire, who has cystic fibrosis, Jane McCabe wrote in *Newsweek* (December 26, 1988): "If you had to choose between saving a cute dog or my equally cute, blond, brown-eyed daughter, whose life would you choose?. . . It's not that I don't love animals, it's that I love Claire more." Ignoring that a single dog experiment could never cure her child's disease, the moral question is whether personal attachment justifies harming others. Since McCabe probably loves her daughter more than other children, would she endorse experimenting on other children to save her child? This, after all, would be a scientifically more productive research strategy than experimenting on nonhuman animals.

Prejudice Is Prejudice

Throughout history, people have set forth systems of rules and laws which excluded others: other clans, other races, other sexes, other religions, etc. To current Western observers, many of these prejudices seem as self-evidently "wrong" as the current exclusion of other species seems obviously "right." Even if we insist on rejecting the universal requirement of ethics, given our propensity for prejudice we should be skeptical of any distinction based on membership of a group, and should seriously question any rules that just happen to benefit us.

It's an undeniable fact that other animals are made of flesh, blood, and bone, just like human beings. They have the same five physiological senses and a range of thought and sensation that is often as developed as many humans. In fact, considering that many have much grander capacities in areas such as sight, smell, and hearing, it's possible that their capacity for joy or pain in certain areas is greater than that of humans (as argued in Jonathan Balcombe's *Pleasurable Kingdom: Animals and the Nature of Feeling Good*). The only clear distinction of membership in the human species is that, with gender, race, and nationality no longer being fashionable prejudices, "human" is the most exclusive group to which most philosophers now pledge allegiance.

Books Referenced

Alinsky, Saul. *Rules for Radicals: A Practical Primer for Realistic Radicals*. New York: Vintage Books, 1972.

Allen, David. *Getting Things Done: The Art of Stress-Free Productivity*. New York: Penguin Books, 2003.

Baur, Gene. *Farm Sanctuary: Changing Hearts and Minds about Animals and Food*. New York: Touchstone, 2008.

Balcombe, Jonathan. *Pleasurable Kingdom: Animals and the Nature of Feeling Good*. New York: Macmillan, 2006.

Bateson, Patrick. "Do Animals Feel Pain?" *New Scientist* 25 April 1992: 30.

Bringas, Ernie. *Going by the Book: Past and Present Tragedies of Biblical Authority*. Charlottesville, Va.: Hampton Roads Publishing Company, Inc., 1995.

Carnegie, Dale. *How to Win Friends and Influence People*. New York: Simon & Schuster, 1998.

Cialdini, Robert. *Influence: The Psychology of Persuasion*. New York: HarperCollins, 2007.

Covey, Steven. *The Seven Habits of Highly Effective People*. New York: Simon & Schuster, 1990.

Druyan, Ann, and Sagan, Carl. *Shadows of Forgotten Ancestors: A Search for Who We Are*. New York: Ballantine Books, 1993.

E. G. Smith Collective. *Animal Ingredients A to Z*. San Francisco: AK Press, 1997.

Eves, Howard. *Mathematical Circles Adieu*. Boston: Prindle, Weber & Schmidt, 1977.

Frey, Raymond G. *Interests and Rights: The Case against Animals*. Oxford: Oxford University Press, 1980.

Gandhi, Mohandas. *Autobiography: The Story of My Experiments with Truth*. New York: Dover Publications, 1983.

_____. "The Moral Basis of Vegetarianism," a speech delivered at a social meeting organized by the London Vegetarian Society, 20 March 1931. See www.ivu.org/news/evu/other/gandhi2.html.

Gilbert, Daniel. *Stumbling on Happiness*. New York: Knopf, 2007.

Gladwell, Malcolm, *Blink: The Power of Thinking without Thinking*. New York: Little, Brown, 2005.

_____. *The Tipping Point: How Little Things Can Make a Big Difference*. New York: Back Bay Books, 2000.

Grandin, Temple. *Animals in Translation: Using the Mysteries of Autism to Decode Animal Behavior*. New York: Simon & Schuster, 2005.

Haidt, Jonathan. *The Happiness Hypothesis: Finding Modern Truth in Ancient Wisdom*. New York: Basic Books, 2006.

Lappé, Francis Moore. *Diet for a Small Planet*. 1975. New York, Ballantine Books, 1985.

Layard, Richard. *Happiness: Lessons from a New Science*. New York: Penguin Press, 2005.

Linzey, Andrew. *Animal Theology*. Champaign, Ill.: University of Illinois Press, 1995.

McCabe, Jane. "Is a Lab Rat's Fate More Poignant than a Child's?" *Newsweek*, 26 December 1988.

Morgenstern, Julie. *Never Check E-mail in the Morning*. New York: Simon & Schuster, 2005.

Pollan, Michael. *The Omnivore's Dilemma: A Natural History of Four Meals*. New York: Penguin, 2007.

Rawls, John. *A Theory of Justice*. Oxford: Oxford University Press, 1999.

Ryder, Richard D. *Painism: A Modern Morality*. Arundel, England: Centaur Press, 2001.

Scully, Matthew. *Dominion: The Power of Man, the Suffering of Animals, and the Call to Mercy*. New York: St. Martin's, 2003.

Singer, Peter. *Animal Liberation*. 1975. New York: HarperCollins Publishers, 2002.

_____. *How Are We To Live? Ethics in an Age of Self-Interest*. Amherst, N.Y.: Prometheus Books, 1995.

_____. *Practical Ethics*. Cambridge: Cambridge University Press, 1993.

_____. "The Singer Solution to World Poverty." *The New York Times Magazine*, 5 September 1999: 60–63.

_____. *Writings on an Ethical Life*. New York: Ecco Press, 2000.

Thoreau, Henry David. *Walden*. Boston: Houghton, Mifflin, and Company, 1897.

Tierney, John. "Go Ahead, Rationalize. Monkeys Do it Too." *The New York Times*, 6 November 2007.

Walker, Alice. *Possessing the Secret of Joy*. New York: Pocket Books, 1993.

Wilford, John. "Almost Human, and Sometimes Smarter." *The New York Times*, 17 April 2007.

Wright, Robert. *The Moral Animal: Why We Are the Way We Are: The New Science of Evolutionary Psychology*. New York: Vintage Books, 1995.

Biographies

Matt Ball, along with Jack Norris, co-founded Vegan Outreach in 1993. Before working full-time as the Executive Director of Vegan Outreach, he was a Department of Energy Global Change Fellow, working in the Department of Forest Ecology at the University of Illinois, and then the Department of Environmental Engineering and the Department of Engineering and Public Policy at Carnegie Mellon. He was also a Research Fellow in the Department of Biology at the University of Pittsburgh.

Bruce Friedrich is Vice President, Policy & Government Affairs for People for the Ethical Treatment of Animals (PETA), the world's largest animal rights organization, where he has worked since 1996. Before coming to PETA, he spent more than six years running a shelter for homeless families in Washington, D.C. and organizing against globalization and war-making.

Lantern's Flashpoint Series explores contemporary social issues in a manner that is direct and challenging. These titles might be incendiary or illuminative. They may set off fireworks or light bulbs, but will galvanize and wake people up.

The Flashpoint Series includes:

An Expensive Way to Make Bad People Worse
An Essay on Prison Reform from an Insider's Perspective

Carbophobia!
The Scary Truth About America's Low-Carb Craze

Aftershock
Confronting Trauma in a Violent World: A Guide for Activists and Their Allies

Strategic Action for Animals
A Handbook on Strategic Movement Building, Organizing, and Activism for Animal Liberation

The Animal Activist's Handbook
Maximizing Our Positive Impact in Today's World

Find these titles at www.lanternbooks.com.